INTEGRATING GIS, REMOTE SENSING, AND MATHEMATICAL MODELLING FOR SURFACE WATER QUALITY MANAGEMENT IN IRRIGATED WATERSHEDS

T0300019

INTEGRATING GIS, REMOTE SENSING,
AND MATHEMATICAL MODELLING
FOR SURFACE WATER QUALITY MANAGEMENT
IN IRRIGATED WATERSHEDS

DISSERTATION

Submitted in fulfillment of the requirements of

the Board for Doctorates of Delft University of Technology

and of the Academic Board of the UNESCO-IHE

Institute for Water Education

for the Degree of DOCTOR

to be defended in public on

Tuesday 24th January, 2012, 12:30 hours

In Delft, the Netherlands

by

Amel Moustafa AZAB

Master of Science in Irrigation and Hydraulics, Faculty of Engineering

Cairo University, Egypt

born in Cairo, Egypt.

This dissertation has been approved by the supervisor:
Prof. Dr. R. K. Price

Committee members:

Chairman	Rector Magnificus TU Delft
Vice-Chairman	Rector UNESCO-IHE
Prof. dr. ir. R. K. Price	UNESCO-IHE/ TU Delft, supervisor
Prof. dr. ir. A. E. Mynett	UNESCO-IHE/ TU Delft
Prof. dr. ir. A. W. Heemink	TU Delft
Prof. dr. ir. N.C. van de Giesen	TU Delft
Prof. dr. M. T. Gaweesh	NWRC, Cairo, Egypt
Dr. Z. Vekerdy	ITC/University of Twente
Prof. dr. M. Menenti	TU Delft, reserve

CRC Press/Balkema is an imprint of the Taylor & Francis Group, an informa business

© 2012, Amel Moustafa Azab

Published by:
CRC Press/Balkema
PO Box 447, 2300 AK Leiden, the Netherlands
e-mail: Pub.NL@taylorandfrancis.com
www.crcpress.com - www.taylorandfrancis.co.uk - www.ba.balkema.nl

ISBN 978-0-415-62115-1 (Taylor & Francis Group)

Summary

Major problems due to specific aspects of human society affect the water quality of rivers, streams and lakes. Problems arise from inadequately treated sewage, poor land use practices, inadequate controls on the discharges of industrial waste waters, incorrect locations of industrial plants, uncontrolled and poor agricultural practices, excessive use of fertilizers, and a lack of integrated watershed management. The effects of these problems threaten ecosystems, endanger public health risks, and intensify erosion and sedimentation, leading to land and water resources degradation. Many of these negative effects arise from environmentally destructive development, a lack of information on the situation regarding water quality and poor public awareness and education on the protection of water resources. Irrigated watersheds are particularly prone to such water quality problems.

Increasing attention is being paid to the management of water resources on a watershed basis, necessitating a cross-disciplinary approach to the definition of problems, data collection and analysis. The assessment of surface water quality on a watershed scale, involves the examination of all activities in the watershed for their possible effects on the existing water bodies. Irrigated watersheds are often complex in their physical nature in that they include interacting irrigation and drainage networks which may be connected to lakes or lagoons. Studying surface water quality problems in such watersheds for better management practices calls for a reassessment and integration of information technology tools designed to support the management process.

Water quality models are considered key elements in understanding water quality problems and are important components in management and decision support systems. Models are now becoming very advanced in describing the dynamics of the aquatic environment and can produce a considerable amount of data, which can be difficult to appreciate. The problem that often arises is selecting the most efficient way of presenting those data in their geographical context. On the other hand, the geographical processing of environmental information is well developed and many advanced Geographical Information Systems (GIS) are now available. Remote sensing techniques have also been shown through different studies to have good potential for mapping and monitoring a number of water quality parameters. Therefore the integration between mathematical modelling, GIS and remote sensing applications could provide a powerful tool for management and decision making process related to surface water quality problems. Also, this integration is an important approach to overcoming the problem of data scarcity in such environments.

The research presented in this study is aimed at contributing to the field of surface water quality management through integrating physically based water quality mathematical models with the spatial capabilities of GIS and the spatial and temporal capabilities of remote sensing in order to develop an integrated water quality management information system that is applicable to irrigated watersheds.

The success of an integrated approach to water quality management in a watershed depends critically on the availability of data. Such data comes in different forms for different levels of users, whether for decision makers and planners or the technical engineers involved in the field. An essential tool used by water quality researchers and managers in developing management plans for rivers, streams and lakes, is mathematical modelling. The integration of various computer technologies and tools, such as GIS and the fast growing technology of remote sensing as a powerful source of data, together with water quality models, gives an even more powerful and efficient management tool, specially when dealing with complicated drainage networks in watersheds.

A general framework for an information system to facilitate the surface water quality management is developed within this research study. This system is based on the integration of hydrodynamic and water quality models with GIS and remote sensing as tools for generating management scenarios for surface water quality in an irrigated watershed. The application is developed for the Edko drainage catchment and shallow lake system in the western part of the Nile Delta, Egypt. The framework includes a hierarchy of modeling tools: a 1D-2D basic hydrodynamic model for a combined shallow lake-drainage system, a detailed 2D hydrodynamic model of the shallow lake, and a 2D water quality and eutrophication models for the lake system. In addition to these modeling tools, remote sensing satellite data are used to calibrate and validate the mathematical water quality models.

As a component of the water quality management information system for the Edko drainage catchment and shallow lake, a 1D-2D hydrodynamic model is developed to understand the basic hydrodynamics of the catchment–lake system. For further analysis of the water quality and eutrophication condition of the lake, a more detailed 2D hydrodynamic-water quality model is developed. This model is based on the 1D-2D modelling concept with a focus on main parameters affecting the lake hydrodynamics such as the wind and tidal time series, the evaporation losses from the water body and evapotranspiration from aquatic vegetation. Reliable water quality models are based on detailed and well structured hydrodynamic models that are capable of describing the physical and hydrodynamic processes of the water system. Excess nutrients loadings that lead to eutrophication, is a common problem in most shallow lakes, especially those linked to agricultural drainage systems. Therefore detailed water quality models for the lake system were needed for inclusion in the framework. 2D hydrodynamic and specific water quality and eutrophication models were developed for the coastal shallow lake Edko.

The water quality model has first a component that simulates the main water quality parameters including the oxygen compounds (BOD, COD, DO), nutrients compounds (NH4, NO3, PO4), the temperature, salinity and the total suspended matter (TSM). The model predicts the basic water quality indicators of the lake system. The second component is the eutrophication screening model for the lake; this is based on the simulation of the chlorophyll-a concentration which is considered an indicator of phytoplankton abundance and biomass in coastal and estuarine waters. A successful application of a model requires a calibration that includes a comparison of the simulated results with measured lake conditions, using different calibration tools. The calibration of the water quality models is done using different techniques to ensure comparable performance of the model to measured data. The model is initially calibrated by adjusting the parameters for selected water quality processes.

A second level of model calibration was for the TSM concentration patterns generated using remote sensing data captured from the analysis and processing of a time series data set of MODIS (MOD09) images specifically for temporal qualitative calibration. A detailed quantitative spatial calibration was done for TSM concentrations based on available *in situ* measurements and the time series MODIS satellite images, by applying TSM analysis algorithms. The SPOT-5 satellite scene was used to extract CHL-a concentrations in order to calibrate the eutrophication model; the results were verified using a MERIS-FR image. The use of remote sensing analysis procedures and data for the calibration and verification of water quality mathematical models in data scarce environments has been shown to provide a valuable and reliable approach.

This study explored the integration of mathematical models and remote sensing methodologies and has shown that there is value in identifying and working with the spatial and temporal variation of the water quality parameters within the lake water body. The calibrated models were used to develop nutrients reduction scenarios for the management of the lake water quality. The reduction scenarios depended on reducing nutrients flow rates from the catchment to the lake. The research succeeded to link the modeling needs with different existing tools for better management, taking into account the practical limitations, and taking a feasible and reliable approach for developing a framework for managing surface water quality in shallow - lake systems with irrigated catchments.

Acknowledgements

The very first person I would like to thank with the deepest appreciation and the most honor which I feel being one of his students is my Promotor Prof. Dr. Roland Price, no words can express my eternal gratitude for his scientific support, encouragement and guidance. I will always be grateful and thankful to him and he will always be in my mind and heart; he taught me how to be a good and confident researcher, how to think beyond research limitations and how to keep working and learning without losing enthusiasm.

The second person who had a great impact on my scientific and technical career during the years of my PhD research, is my dear and sincere supervisor Dr. Zoltan Vekerdy from ITC-University of Twente, he was always there for me, I am grateful for his time and effort, I thank him for all the valuable time that I spent at ITC-Enschede and the vast knowledge that I gained from there, all the long discussions we had on remote sensing and the ideas he gave me were not only reflected in my thesis, but they shaped my future interest as a researcher, he taught me that every simple scientific idea is valuable, if it is not a step taken today it could be a step forward taken tomorrow. Thank you Dr. Vekerdy and thanks to ITC for the remarkable scientific support to me and providing me with all the needed remote sensing data and satellite images needed in my research application.

My special thanks also goes for Dr. Ioana Popoescu, my dear supervisor from Hydroinformatics Department of UNESCO-IHE, I greatly appreciate her help, time and support in developing the mathematical modelling part of the research, I will always remember her guidance and encouragement every time we met to discuss further steps in modelling.

My deepest thanks and appreciation are for my sincere supervisor from Egypt, Prof. Dr. Moustafa Gaweesh, his support from the first step till the last step of my research was remarkable, and his amazing commitment as a supervisor was always supporting me to work hard under any circumstances. I will always remember his help and continuous follow up during data collection in Egypt and how keen he was to assist me to get access to all available data sources. I feel blessed to learn from him and his scientific, technical and life experiences.

My great appreciation and true thanks is for PoWER project of UNESCO-IHE (Partnership for Water Education and Research), the project that supported me financially to conduct this PhD research. I can't forget the support of the PoWER project first Director, Dr. Atem Ramsundersingh, it was Dr. Atem who convinced me to take this innovative research opportunity to carry out my PhD research through a collaborative research project under PoWER, from that moment innovation and collaboration became part of my technical career, thanks Atem. I was further blessed by the continued support and encouragement of Prof. Dr. Jetze Huen the current Director of PoWER, who was eager to continue the mission of PoWER and to take it to another level of success, Thanks Prof. Jetze for you encouraging spirit.

I am also thankful to the Hydraulics Research Institute (HRI), my home Institute in Egypt, My appreciation goes to Eng. Ibrahim El Desouki, the former acting Director of HRI for supporting me to arrange for my field work and survey mission to lake Edko. I am grateful to all the HRI survey team who joined me during my several field work missions in the lake, special thanks are for the survey team technical staff Mr. Hassan, Mr. Mahmoud Swilem and all the team members. I acknowledge also all the assisting team at the Central Laboratories for Environmental Quality Monitoring (CLEQM) of the Egyptian National Water Research Center, who helped me in water quality sampling and analysis during and after field work, appreciation for Dr. Mohamed Mokhtar, Eng. Wahba and all CLEQM technical staff.

All the gratefulness, admiration and sincere thanks are to my second family, the Nile Basin Capacity Building Network (NBCBN), UNESCO-IHE Management team and Cairo Secretariat office team. You were all the enchanting supporters behind any success I ever achieved in my technical career in the last ten years. I am grateful to Ir. Jan Luijendijk and Prof. Petru Boeriu, i wouldn't have gone through both tough challenges, practical and scientific careers, unless you supported and encouraged me, my deepest thanks to both of you, you are so great, you will always have a special place in my heart!. My thanks and appreciation goes to Drs. Carel Keuls, UNESCO-IHE NBCBN project advisor, thanks so much for your true support for the last year and your sincere help in the Dutch translation of my PhD summary and Propositions. My NBCBN colleagues and true friends; the Secretariat office team, my colleague and sincere friend and brother Dr. El-Sayed Diwedar, I am and will always be grateful for your sincere support and kind spirit, my dear colleagues and true friends Eng. Hend Haider and Mrs. Nashwa Nader, no words can give you your true credit, thank you for being always beside me. Special thanks also goes to my colleague and research fellow in the Wetlands Management Research Group of NBCBN, Mr. Mohamed Saeed, GIS expert, your ideas and support in the geo-database development are highly appreciated.

My appreciation goes to all my UNESCO-IHE colleagues that I have known during the years of my research and who were true friends during my stay in the Netherlands, special thanks to Mohamed Bahgat and Sherif Megahed, and to my PhD colleagues Aya Lamie, Dima, Sherif Waly and Marmar Badr, I will always remember you.

Last but not least, no words in the world can express my thanks, admiration, love and appreciation to those who were behind me and beside me all the time, to the most kind hearts in my world to my family; to the soul of my late beloved father and to my great beloved mother, thank you for your unconditional continuous support and love for me, I wouldn't make it without you. To my life partner, my beloved husband Ahmed and my adorable kids Nour and Mazen, thank you all for your kind hearts, your true and sincere love and your confidence in me, you were my supporting strength in all the stages of my research.

Finally, I thank GOD, his support that is above all supports, I always put my faith in GOD and GOD never let me down.

Table of Contents

1. INTRODUCTION

The assessment of surface water quality on a watershed scale, involves the examination of all activities in the watershed for their possible effects on the existing water bodies. Agricultural irrigated watersheds are of complex physical nature in that they include interacting irrigation and drainage networks which may be connected to lakes or lagoons. Studying surface water quality problems in such watersheds for better management practices calls for a reassessment and integration of information technology tools designed to support the management process. Therefore the integration between mathematical modelling, GIS and remote sensing applications could provide a powerful tool for management and decision making process related to surface water quality problems. The present research aims to contribute to the field of surface water quality management through integrating physically based water quality mathematical models with the spatial capabilities of GIS and the spatial and temporal capabilities of remote sensing to develop an integrated water quality management information system that is applicable to irrigated watersheds. Edko drainage catchment and shallow lake system in the western Egyptian Nile Delta is chosen as a pilot watershed for application of the proposed system.

1.1. BACKGROUND

Many developing countries including the Nile Basin Countries face threats to the security of their water resources. Fulfilling the increasing needs for water supply, irrigation and hydropower results in water scarcity, a serious decline in water quality and growing environmental and social concerns. With the consequences due to floods and droughts aggravated by climate change, the need for sustainable management of available water resources becomes a key issue for the future development of these countries.

The surface water quality of lakes, reservoirs, rivers and drainage channels can vary in space and time according to natural morphological, hydrological, chemical, biological and sedimentation processes. Pollution of natural bodies of surface water is widespread because of human activities, such as disposal of sewage and industrial wastes, land clearance, deforestation, use of pesticides, mining, and hydroelectric developments. However, clean water is essential to human survival as well as to aquatic life. Much surface water is used for irrigation, with lesser amounts for municipal, industrial, and recreational purposes: only 6% of all inland water is used for domestic consumption. An estimated 75% of the population of developing nations lacks adequate sanitary facilities, and solid waste is commonly dumped into the nearest body of flowing water. Pathogens such as bacteria, viruses and parasites make these waste materials among the world's most dangerous environmental pollutants: water-borne diseases are estimated to cause about 25,000 deaths daily worldwide (World Bank, 2000 (World Bank 2000)).

 Pollution sources affecting surface water bodies and contributing to the increasing deterioration of water quality in general fall into two categories: point sources and non-point sources (NPS). The point sources include discrete flows of polluted water that enter a stream through a pipe or channel, for example, the effluent from a sewage treatment plant. Point sources are often associated with industries or municipalities. Non-point sources are diffuse contributions that occur over a wide area and are usually associated with land uses such as agricultural cultivation, livestock grazing and forest management practices. These sources generally enter streams as overland flow (i.e. urban runoff), groundwater flow, or flow from small tributaries (McCutcheon, 1990). Abatement efforts to address non-point sources include the identification and implementation of improved land use practices in rural areas involving, for example, agricultural, forestry and road construction activities (Brooks et al., 1997).

In order to manage surface water quality for better protection and remediation, rather than focusing just on water quality of a lake or stream and point or non-point sources of pollution separately, all activities in the surrounding area are examined for their possible effects on the water body. In other words, for better assessment of the problems related to surface water quality, there should be a broad scope of study based on a watershed or an integrated catchment approach (Haith and Tubbs, 2003).

Following the integrated watershed management approach in terms of water quality, the availability of data in different forms and for different levels of users, whether the decision makers and planners or the technical engineers involved in the field, is an issue of great importance. An essential tool, which can be used by water quality researchers and managers in developing management plans for rivers, streams and lakes, is the mathematical modelling of water quality. Integrating different computer technologies and tools, such as GIS and the fast growing technology of remote sensing (as a powerful source of data acquisition), with water quality models, gives an even more powerful and efficient management tool, especially when dealing with complicated surface networks in watersheds. Not only does a GIS enable the user to collect, store, manipulate, analyse, and display spatially referenced data, it also provides an interactive environment to construct models and makes the models easier to use (Goodchild, 1996).

Remote sensing techniques have also shown through different studies good potential for monitoring, mapping and estimating a number of water quality parameters. The reason that these techniques are not yet used widely and efficiently at their full potential can partly be explained by poor information facilities and a poor infrastructure for acquisition, processing, archiving and distributing remote sensing data. It is expected that improvements in these facilities will lead to a considerable increase in the use of remote sensing for water quality applications. With advances using remote sensing for data acquisition and the integration of remotely sensed data with GIS applications and modelling tools, the development of an integrated and improved system for management can be achieved for better assessment and protection of surface water quality.

1.2. THE NATURE AND CHARACTERISTICS OF IRRIGATED WATERSHEDS

It is possible to outline the land area that delivers surface and subsurface water to a particular stream or river by determining the slope of the land and delineating the likely path of water towards one stream or another. The land area that supplies water to a specific stream or river is known as its watershed. Put simply, a watershed is the area of land that supplies water to a lake or river system. Watersheds can be large or small. Every stream, tributary, or river has an associated watershed, and small watersheds aggregate together to become larger watersheds. The connectivity of the stream system is the primary reason why aquatic assessments need to be done at the watershed level. Connectivity refers to the physical connection between tributaries and the river, between surface water and groundwater, and between wetlands and these water sources. Because water moves by definition downstream in a watershed under gravity, any activity that affects the water quality, quantity, or rate of movement at one location can change the characteristics of the watershed at locations downstream. Figure (1-1) shows a typical watershed.

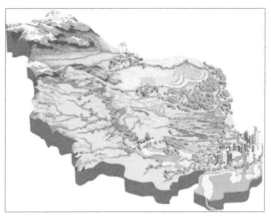

Figure (1-1): A typical watershed

To understand better the interaction between human activities and the environment, it is important to utilize natural landscape features such as watersheds to define the basic unit of analysis and to take all components into consideration, rather than to use artificial units such as political boundaries. Increasingly, watershed-level analysis is becoming the standard method for assessing the vulnerability of aquatic systems. Utilizing a GIS and information derived from remotely sensed data in conjunction with other environmental data sets, watersheds can be examined as the sum of interrelated components. For example, the total area of the watershed, hydrologic connectivity of the surface water, amount of cropped land, precipitation, soils and other environmental data can all be combined to assess the vulnerability of the watershed as a whole, or individual point source intakes within the hydrologic system.

An irrigated agricultural watershed could be considered as a complex type of watershed, in addressing problems related to water quality. The complexity of such watersheds comes from the fact that they usually comprise intensive networks of irrigation and drainage channels. These networks are connected to open surface water bodies such as lakes, lagoons, estuaries or directly connected to the open sea. Furthermore, the land use in such watersheds can be mixed: the main land use could be agriculture, but other various categories may include fisheries, industrial areas, urban areas and discrete rural communities. Therefore, pollution problems of surface waters in such types of watersheds are likely to be complex due to the presence of combined sources of pollution in the form of point and non-point sources based on different types of land use.

1.3. WATER QUALITY PROBLEMS IN CONNECTED CATCHMENT-SHALLOW LAKE SYSTEMS

One of the most important problems associated with irrigated watersheds is the non- point source of pollution resulting from agricultural practices, introducing various pollutants to the drainage network such as phosphorus, nitrogen, metals, pathogens, sediment, pesticides, salt, trace elements (e.g. selenium). This is in addition to the pollution load from domestic untreated wastewater, fisheries waste loads, and other point sources of pollution from inadequately treated industrial and domestic wastewater. In irrigated agriculture, the conservation of water by the efficient conjunctive use of irrigation and drainage water may provide an economically sound solution to the problem of irrigation water deficiency in countries with limited water resources. Reuse of drainage water either directly or by mixing with irrigation water is one of the agricultural practices in irrigated watersheds. However, this practice has its limitations and drawbacks. Deterioration in of the quality of drainage water by different pollution sources has direct negative effect on irrigation water quality, agricultural crops and local farmers. Coastal deltaic regions with rare rainfall are usually a good example of such types of watersheds.

The Nile Delta region in Egypt, could be referred to here as a typical example, the region is divided into a number of irrigated agricultural watersheds that comprises drainage catchments connected to coastal shallow lakes, which act as drainage system outlets to the sea. All human activities that take place in the catchments, has its effect on the whole watershed. Irrigation in the Nile Delta depends on drainage water reuse, as a second important source after surface irrigation. Rainfall on the Mediterranean coastal strip decreases eastward from 200 mm/year at Alexandria to 75 mm/year at Port Said. It also declines inland to about 25 mm/year near Cairo. Rainfall occurs only in the winter season in the form of scattered showers. Therefore, it cannot be considered a dependable source of water. Therefore the surface water pollution sources are mainly originating from agricultural drainage water in addition to industrial and untreated domestic wastes. The pollutants are transferred from the drainage network to the coastal shallow lakes. These pollutants are mixed with another source which is the wastewater from aquaculture or fish ponds that usually surrounds coastal lakes. The ecosystems of such lakes usually face eutrophication problems due to excess nutrient loadings into the lakes. Also, there are usually unbalanced quantities and species of aquatic vegetation due to existing pollution problems.

The pollution problems in such watersheds, especially in developing countries including the Nile Delta catchments, are increasing due to the lack of formal sanitation and treatment facilities, population growth, continuous urbanization and the lack of public awareness on the extent of the problems and the importance of pollution prevention. In these watersheds, which are subject to different pollution sources, there is an urgent need for effective management of the surface water quality. Management tools are needed in order to ensure compliance of surface waters with the different water quality objectives for various uses. Environmental management problems associated with water quality in such watersheds are inherently complex and difficult to analyze due to the interaction of physical, chemical and biological processes and the consequences of anthropogenic activities. Further, finding good management alternatives becomes exceedingly difficult due to conflicting issues such as cost, environmental impact and equity that need simultaneous consideration. The solution of these problems is compounded by a very complicated system that includes different types of water bodies, categories of land use and sources and types of pollutants. Therefore, the solution requires an integrated approach to the modeling, analysis and management of such types of watersheds in order to manage their surface water quality.

1.4. HYDROINFORMATICS APPLIED TO WATER QUALITY MANAGEMENT

Classical hydro engineering (hydraulics, hydrology and related research), linked to meteorology and water quality, usually deals with just one aspect of the total problem. As a consequence, the results of hydraulic research, as well as modelling software, need to be integrated into larger systems to reach a holistic approach to solving real life problems. Such problems have to be seen in the context of a more comprehensive exchange of information concerning real world water-based issues and the interests and intentions of their various stakeholders. Here, the role of hydroinformatics becomes more obvious and important. Hydroinformatics is a socio-technology built around developments and applications of systems which are, for their users, objective. A tool is objective if the users are involved in its definition, if they can easily understand the results and use them, if they have the possibility to input their own hypotheses into the system and see the consequences as well as to show these to other stakeholders. In the water sector, the society's needs and requirements are real, and linked to real life problems. The more that society becomes aware that it depends upon water, the more it understands that water is central to sustained development not just at the local but at the national level. Solutions to such problems go beyond traditional hydraulics and hydrology. Hydroinformatics changes the way in which hydraulics, hydrology and water resources studies are applied in society. In solving real life problems, hydroinformatics assists in developing a one-to-one mapping of the real world onto a virtual parallel world created by applying the information and communication tools. The virtual world of models and different information management tools applied to specific problem areas, are translated to society and significant stakeholders in the form of proposed solution scenarios for the prescribed water problems. This is done through the procedural world involving the different levels of professionals who implement and run the virtual world and interpret its results into reliable information and conclusions. In managing water quality problems this framework of different worlds or environments is very appropriate: the real world problem area, the virtual representation, the procedural world and the societal world are considered the main building blocks of this framework. Figure (1-2) shows the framework for solving real world water related problems applying hydroinformatics tools.

If we consider surface water quality problems in irrigated watersheds and apply this framework, the *real world* is represented by the problem area of interest, i.e. the watershed including its catchment area and all associated water bodies such as drains and lakes. *The virtual world* in parallel to the problem area is represented by the developed mathematical models and all tools linked to it, such as GIS and remote sensing. The interface between the real world and the virtual representation includes the conceptualization of the problems and issues, the selection of solution methodologies, the formulation and collection of data sets including hydrodynamic related data, geometrical parameters, water quality processes to be represented, water quality parameters, the selection of modelling tools,…etc. The analysis of the water quality problem takes place in the virtual world component. The third component, namely the *procedural world,* involves professionals, i.e. the near-end users of the virtual world, dealing with the virtual system according to prescribed or ad-hoc procedures. The interface between the virtual world and the procedural world includes the application and implementation of models, interpretation of models results and continuous calibration and enhancement of models and tools based on feedback from the far-end users.

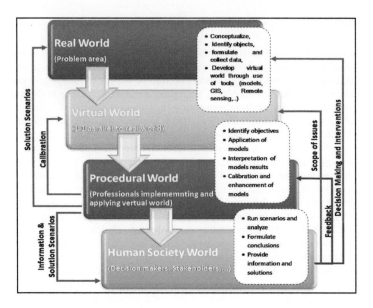

Figure (1-2): The framework for solving real world water related problems

The near-end users identify the target output objectives of the virtual world, run the analysis scenarios and formulate objective conclusions. The last building block is the *human society world*, or the far-end users, involving the decision makers and different stakeholders. The interface between this world and the procedural world includes the development of reliable and efficient information in the form of results from the virtual world, the development of solution scenarios based on model results, and the receipt and application of feed-back enhancements to the virtual world. The human society world is responsible for the safe, sustainable and efficient interventions that take place in the real world to solve the water quality related problems. In total, hydroinformatics is the implementation tool of integrated water resources management.

1.5. THE IMPORTANCE OF DATA AVAILABILITY

In the context of surface water quality management, data availability is a crucial issue to be considered. Sustainable management of water quality involves dealing with different categories and huge amounts of data. The availability of the right data sets, which are used to develop the appropriate information needed for a water quality decision support system (DSS), is one of the problems that face us in the management process. The over-arching problem of data programs (monitoring and data use) was summarized by Ongley (1997) as:

> "… a common observation amongst water quality professionals is that many water quality programs, especially in developing countries, collect the wrong parameters, from the wrong places, using the wrong substrates and at inappropriate sampling frequencies, and produce data that are often quite unreliable; the data are not assessed or evaluated, and are not sufficiently connected to realistic and meaningful program, legal or management objectives. This is not the fault of developing countries; more often it results from inappropriate technology transfer and an assumption by recipients and donors that the data paradigm developed by western countries is appropriate in developing countries. "

In developing countries, data collection programmes are most of the time data driven rather than needs or objective driven. Regrettably, many countries including developed countries, entrust data programmes to agencies that have data-collection as their primary mandate, with the result that water quality data programmes exhibit a high degree of inertia and for which there are few identified users of the data. The consequence has been the realization that these mainly chemistry-focused programmes are expensive, they focus on data production rather than on data use, and collect more data than is necessary. They often do not reflect the types of data that managers need, and can often be replaced by cheaper and more effective methods. The outcome in Canada and the United States, as an example, has been a substantial shrinkage of conventional water quality data programmes and an expansion of alternative approaches. Nowadays these are expensive, and the often ineffective chemistry-focused approach is the one now being adopted by most developing countries and is being recommended by international and multilateral organizations (Ongley, 2000). This is one of the main reasons that most of the collected water quality data in developing countries is not used efficiently in real analysis and management plans. Water quality analysis is mainly done through trends and the development of simple indicators, but detailed water quality modelling is not yet effectively taken into consideration in the real implementation of existing DSS or management plans.

For developing appropriate water quality management tools there should be an effective objectively-oriented monitoring system that serves the real needs of the management sector, rather than just collecting huge amounts of data that might not be needed in managing real water quality problems. Also, integrating or complementing the monitoring systems or data collection programmes with other effective types of data is an important step in developing new approaches for managing water quality. Such types of data may include digital maps and remote sensing information based on in-situ measurements and satellite images. This type of integrated approaches is considered a step forward in developing reliable DSSs for surface water quality management.

1.6. THE NEED FOR INTEGRATED MANAGEMENT TOOLS

In view of the physical complexity of the watersheds and the need to deal with large and various amounts of data in integrated and interacting watersheds and water systems, there is a requirement for reliable and powerful analysis tools. To study water quality problems associated with such systems, access to information for managing such complicated problems is an essential and fundamental need. The use of Geographic Information Systems in combination with remote sensing is considered the most suitable method that enables realistic, direct and reliable access to information. It is also very important to understand the different surface water problems. This understanding indeed becomes an important part of the solution. Modelling is the primary tool to support this task (Radwan, 2002). In particular, proper management of a complicated network of canals and drains requires different levels of modelling pollutants and their transport, especially the non-point sources. An extensive knowledge is required of the different categories and the collection of various amounts of data including: land use, soil properties, land slope, agricultural activities, socioeconomic background, in addition to the different types of data related to the network itself, and the existing connection and relation between the drainage network water and other water bodies.

The main challenge in building effective information systems for integrated water quality management is the integration of dynamic models with the capabilities of GIS and Remote Sensing. The GIS can provide a common framework of reference for the various tools and models addressing a range of problems in river basin management as a whole. In a multi-tool framework, it can also provide a common interface to the various functions of an information and decision support system for integrated water quality management. This interface has to translate the available data and model functionality into information that can directly support the decision making processes (Fedra, 1996). Remote sensing is considered a very strong data acquisition tool, which has great possibilities regarding the determination of some water quality parameters and detailed information on the lakes. In addition, to data acquisition, remote sensing is considered a powerful tool in estimating some water quality parameters of great concern especially in studying eutrophication of shallow lakes such as total suspended matter (TSM) and CHL-aorophyll-a (CHL-a).

Geographic Information Systems are becoming more useful in modelling water quality because they can readily incorporate spatially varying data. There are many instances where GISs have been incorporated into modelling efforts. Two basic ways that they have been used are: (1) as a method for deriving input for external models, and (2) as a stand-alone data model. The coupling of GIS to a water quality model is a marriage designed to address the problem of spatially simulating non-point sources of pollutants at field, basin, region, and global scales. A GIS provides both the means of organizing and manipulating spatial data and of creating visual displays of geo-referenced data. Progress has been made in coupling GIS with water quality models on the basis of either a loose or close coupling strategy.

In the design and development process of decision support systems for integrated water resources management and planning, water quality issues are a major component that should be taken into account. Water quality mathematical models are considered essential components of the computational framework of such a DSS. Modelling, in the decision making process, provides the answers to particular questions associated with environmental problems. Several studies have been carried out both on building a standalone DSS for water quality management and on developing the appropriate computational frameworks of DSS including water quality modelling tools. These water quality DSSs depend on structured databases and computational frameworks. The computational frameworks comprise water quality modeling tools of different properties and approaches, which depend on many factors such as the particular water quality modeling problems, scale of application, parameters of concern and management objectives.

Remotely sensed data and information derived from them have a wide range of application in hydrology and water resources management (Shultz, 1988). Remote sensing and its associated image processing technology provide access to spatial and temporal information at watershed or at regional continental and global scales. Effective utilization of this large data volume is dependant upon the existence of an efficient, geographic handling and processing system that will transform these data into usable information. A Geographic Information System is a major tool for handling this spatial data (Mattikalli and Engman, 2000).

Remotely sensed data can be best utilized if they are incorporated in a GIS that is designed to accept large volumes of spatial data. Applications of GIS and remote sensing have mainly concentrated on non-point sources (NPS) of pollutants. This is because remotely sensed data products such as land-use/land cover could be directly utilized in NPS modelling.

Monitoring and assessing water in streams, reservoirs, lakes, estuaries and oceans are critical aspects for managing and improving the quality of the environment. Classical techniques for measuring indicators of water quality involve *in situ* measurements and/or the collection of water samples for laboratory analysis. Although these technologies give accurate measurements for a point in time and space, they are time consuming to implement, expensive, and do not give either the spatial or temporal view of water quality for an individual water body or multiple water bodies across a landscape. Remote sensing of indicators of water quality and its interpretation using GIS, offers the potential of relatively inexpensive, frequent, and synoptic measurements using aircraft and/or satellites (Ritchie and Schieb, 2000). Suspended sediment, CHL-aorophyll (algae), (humus), oil, and temperature are water quality indicators that can change the spectral and thermal properties of surface water and are the most readily measured indicators by remote sensing techniques. Substances (i.e. chemicals) that do not change the optical and thermal conditions of surface waters can only be inferred by modelling using other surrogate properties (i.e. suspended sediments, CHL-aorophylls), which may have responded to an input or reduction of chemicals. Application of remote sensing in measuring suspended sediments, CHL-aorophylls and temperature is widely used.

1.7. PROBLEM STATEMENT

For better surface water quality management of complex irrigated watersheds, there is a need to identify the different components of a watershed, the categories of land use, and the interaction between various connected water bodies. Understanding the hydrodynamics of the water bodies and the different forces affecting them, leads to a better understanding of the water quality problems associated with these water bodies. This truly reflects the requirement for and the expectations made of effective tools in water quality management. Using existing tools for the management of surface water quality in a new and integrated way could be a better approach for finding appropriate solutions for pollution problems. Water quality models, as one set of these tools, are considered key elements in understanding water quality problems and are therefore main components in management and decision support systems. Models describing the dynamics of the aquatic environment are becoming more sophisticated in that they include better and detailed interpretation of water quality processes. In this way they can produce a considerable amount of data, which can be difficult to appreciate; the problem that often arises is how to present these data in their geographical context in the most efficient way. This introduces a reassessment and integration of information technology tools designed to support the management process, and raises the role of the integration of the modelling with GIS and remote sensing.

The advances made in water quality modeling using GIS, Remote Sensing and the importance of management information systems (MIS) and decision support systems (DSS) in the management process are increasingly being recognised. The research reported in this thesis makes use of the integration of GIS capabilities and remote sensing facilities with water quality modeling to develop a computational framework for a DSS concerning surface water quality management in complex irrigated watersheds. The research also explores advances in these tools to solve particular water quality problems. The main scope of the integration is to understand better the water quality of different types of connected water bodies in irrigated watersheds. This integration provides a better and clearer assessment of the water quality problems and helps in developing remedial management actions for future protection of the environment.

1.8. RESEARCH OBJECTIVES

The general objective of this research work is to develop a practically applicable Surface Water Quality Management Information System WQMIS (including the assessment, modelling and management components) that is applied at a watershed scale. This system answers both planning and technical questions of water quality managers, decision makers, and those of technical engineers working on the sampling, monitoring, analysis and modelling of water quality parameters. This system is developed as an integrated computational framework and decision support system for surface water quality management in complex watersheds, and focuses on the flow of information regardless of the user's level.

The proposed system comprises four main components:

The First component is a comprehensive *surface water quality geo-database* including different watershed components, namely the catchment and lake systems. This data base is intended to be the core of the developed information system. It includes all the categories and types of data that are used as either inputs to the modelling components or outputs from models used for DSS. This geo-database is also designed to be accessible by different levels of users based on the level of data needed and the application. The geo-database incorporates all types of data needed for the development and operation of water quality models. It includes, hydrographic survey data, GIS layers of the system under investigation, spatially referenced water quality data, satellite images and processed images, spectral water quality measurements,...etc.

The Second component is the ***Remote sensing tool***. In this developed information system, the role of remote sensing is emphasised for both data acquisition and as a key component upon which the DSS is based. The remote sensing data is applied in all steps of developing the DSS computational framework: from the data acquisition phase for building the geo-database, to the analysis of measured field data and processing of images to developing the algorithms for extraction of concentration maps of TSM and CHL-a and finally for using the developed data in the calibration of water quality models. Therefore remote sensing and associated analysis techniques for water quality are essential tools in this study and the modelling component is dependent on them. We can summarise by stating that the remote sensing tool in this system is considered a critical tool for decision support.

The Third component is a consistent hierarchy of both simple and complex ***modelling tools for water quality pollutant transport and concentration analysis***. This modelling tool specifically comprises the following modules:

- The first module is a simple 1D-2D catchment-lake model for the surface network and connected lake system, to understand the drainage-lake system hydrodynamics and to investigate the environmental problems related to pollution sources (domestic, industrial, agricultural run-off). This modelling tool is considered the base model and it aims at describing the hydrodynamic behaviour of the catchment-lake system and the effect of the catchment hydrology on the lake hydrodynamics.

- The second module is a detailed 2D shallow lake model, which is considered to be the main model of the lake system, which is at the outlet of the drainage catchment. The lake model gives a detailed overview of the lake system hydrodynamics and simulates the spatial and temporal variations of water quality parameters in the lake including the eutrophication parameters. This model is based on the outputs from the catchment model. The water quality model is divided into a model for the basic indicators and an eutrophication screening model. The two models give a complete overview of the water quality condition within the lake, and reflect the conditions within the watershed.

The Fourth component is a water quality DSS tool which is the water quality and eutrophication modelling components of the shallow lake which are used for different management approaches and scenarios. This management tool depends on developing a basic uncertainty analysis to quantify the level of confidence in the modelling system as the first step in conducting a risk assessment.

The specific research objectives of this study are:
- To develop an integrated geo-database to be used as a main data source for watershed surface water quality modelling and management.
- To develop the computational framework for modelling surface water quality model for a connected drainage network-lake system.
- To assess the potential for eutrophication of the end coastal lake, which is the outlet of the watershed to the sea.
- To calibrate the water quality modelling tool using remote sensing data.
- To identify and develop the most critical surface drainage water quality indicators
- To simulate and predict the temporal and spatial variation of pollution concentrations of CHL-aorophyll-a (CHL-a) and total suspended solids (TSS) using integrated mathematical model and remote sensing data.
- To formulate and explore different modelling scenarios for studying the impact of alternative water quality management practices in the selected drainage catchment and their effect on the environmental condition of the coastal lake as an important component of the watershed.

1.9. THESIS OUTLINE

This thesis consists of nine chapters. A brief overview of the content of the following chapters is presented here.

Chapter 2 presents an overview of water quality management in connected catchment-shallow lake systems. A description of the common features and characteristics of connected catchment-shallow lake systems is given including the physical characteristics of water bodies, land and water use. This is followed by a description of the different pollution sources and environmental risks facing these systems. The chapter ends by highlighting the trends and difficulties facing efficient water quality management in these connected systems, by exposing the different management issues and by identifying the relevant tools.

Chapter 3 presents a description of the existing tools for surface water quality management and explains the importance of their integration. The application of the relevant tools including remote sensing, GIS and mathematical modelling are discussed. The integration of tools is explored with reference to the applications of management information systems and their different levels of users. The integration of the different tools into decision support systems for water quality management is highlighted at the end of the chapter.

Chapter 4 introduces the framework of the proposed water quality management information system (WQMIS), it shows the links between the different components of this framework; the geo-database, the hierarchy of hydrodynamic and water quality models and integrated modelling and remote sensing techniques. It guides the reader to get a general background on the selected modelling tools and procedures and how are these models integrated with remote sensing techniques for better understanding of the hydrodynamics and water quality processes of surface water within the study area.

Chapter 5 introduces the study area and its characteristics as an irrigated watershed in the Nile Delta region. It is considered an important chapter to the rest of the thesis as it explains the components of the irrigated watershed, the main water quality problems within the catchment and lake system. It details all the categories of data collected and needed to build the WQMIS and the different data acquisition methods including field work. The chapter gives a detailed overview of preliminary water quality analysis based on collected datasets.

Chapter 6 deals with the water balance and pollutants loads estimates within the lake water system. The chapter gives an estimate of the lake water budget based on measured discharges at the different lake boundaries (drains outlets and exit channel to the sea). It includes a summary on the types of fertilizers used in agriculture being the main sources of nutrients loadings into the lake. The last part of this chapter focuses on the analysis of nutrients loads inputs to the lake based on measured historical data in comparison with literature published values.

Chapter 7 is a key important chapter in the research study since it focuses on the main components of the WQMIS which is the set of hierarchical 1D-2D hydrodynamic, water quality and eutrophiocation models. The chapter explains the development procedures of different models and it highlights the relation between these models and the uses of different modelling component. It ends by the first level of calibration of important water quality indicators (TSM and CHL-a) as a first step towards introducing the importance of Remote Sensing techniques in the models calibration. This chapter is considered an important contribution to integrating different mathematiclas modelling tools for better understanding of complex water systems.

Chapter 8 is the bridging chapter between mathematical modelling and remote sensing. It focuses on the different methodologies and remote sensing techniques used for extracting water quality parameters from satellite images data, and how this driven data is used to complement the lack of measurements for water quality models calibration. The chapter is considered an added contribution to the calibration and validation procedures of water quality and eutrophication mathematical models. It shows the importance of integrating the different analysis tools for getting better understanding and developing better and more reliable management tools.

Chapter 9 links the developed WQMIS to the decision process by developing decision support information from the modelling results. Based on the system outputs, the chapter focuses on the methodology for developing planning management scenarios for abating the increased pollution loads entering the downstream end lakes of physically complex watersheds. The chapter highlights the prediction scenarios for eutrophication management of a shallow lake system by reduction of nutrients loads entering the lake system.

Chapter 10 highlights the general and final conclusions of the research study; it shows the innovative contributions to surface water quality management, though integration of different tools and techniques involving mathematical modelling, GIS and remote sensing. The chapter ends by a set of recommendations for future research work.

2. WATER QUALITY MANAGEMENT IN CONNECTED CATCHMENT- SHALLOW LAKE SYSTEMS

Physically complex irrigated watersheds in general may have some common features such as similar physical components including the drainage catchment, the downstream connection to lakes, lagoons or a direct connection to sea or ocean. The geographic location, hydrology, land use and associated pollution problems are considered with the specific characteristics of a watershed. Water quality models are considered key elements in understanding water quality problems and are main components in management and decision support systems. Understanding the different water quality processes within the connected water systems is important to develop the proper frameworks for management. Also understanding the gaps and problems associated with data availability is important to select the suitable computational frameworks and their components.

2.1. MAJOR CHARACTERISTICS OF WATER BODIES

According to Meybeck *et,al.*, 1996, water bodies can be fully characterised by the insights of three major fields of study: hydrology, physico-chemistry, and biology. The following sections summarize these major fields.

Hydrodynamic Features

All freshwater bodies are inter-connected through the hydrological cycle. Inland freshwaters, which appear in the form of rivers, lakes or groundwater, are closely inter-connected and may influence each other directly, or through intermediate stages, and each water body has its specific hydrodynamic properties.

Rivers are characterised by uni-directional flow with a relatively high, average water velocity ranging from 0.1 to 1 m/s. The river flow is highly variable in time, depending on the climatic situation and the drainage pattern. In general, thorough and continuous vertical mixing is achieved in rivers due to the prevailing currents and turbulence. Lateral mixing may take place only over considerable distances downstream of major confluences.

Lakes are characterised by a low, average current velocity of 0.001 to 0.01 m /s (surface values). Therefore, water residence times, ranging from one month to several years, are often used to quantify mass movements of material. Currents within lakes are multi-directional. Many lakes have alternating periods of stratification and vertical mixing, the periodicity of which is regulated by climatic conditions and lake depth.

Groundwaters are characterised by a rather steady flow pattern in terms of direction and velocity. The average flow velocities commonly found in aquifers range from 10^{-10} to 10^{-3} m /s and are largely governed by the porosity and permeability of the geological material. As a consequence mixing is rather poor and, depending on local hydrogeological features, the ground-water dynamics can be highly diverse.

The hydrodynamic characteristics of each type of water body are highly dependent on the size of the water body and on the climatic conditions in the drainage basin. The governing factor for rivers is their hydrological regime, i.e. their discharge variability. Lakes are classified by their water residence time and their thermal regime resulting in varying stratification patterns. Although some reservoirs share many features in common with lakes, others have characteristics which are specific to the origin of the reservoir. One feature common to most reservoirs is the deliberate management of the inputs and/or outputs of water for specific purposes. Groundwater aquifers greatly depend upon their recharge regime, i.e. infiltration through the unsaturated zone, which allows for the renewal of the groundwater body.

There are several transitional forms of water bodies which demonstrate features of more than one of the three basic types described above and are characterised by a particular combination of hydrodynamic features.

Reservoirs are characterised by features which are intermediate between rivers and lakes. They can range from large-scale impoundments, such as Lake Nasser, to small dammed rivers with a seasonal pattern of operation and water level fluctuations. The hydrodynamics of reservoirs are greatly influenced by their operational management regime.

Flood plains constitute an intermediate state between rivers and lakes with a distinct seasonal variability pattern. Their hydrodynamics are, however, determined by the river flow regime.

Marshes are characterised by the dual features of lakes and phreatic aquifers. Their hydrodynamics are relatively complex.

Alluvial and karstic aquifers are intermediate between rivers and ground-waters. They differ, generally, in their flow regime which is rather slow for alluvial and very rapid for karstic aquifers. The latter are often referred to as underground rivers.

The knowledge and understanding of the hydrodynamics of the water bodies is a crucial requirement before a water quality monitoring or modeling tool can be established. Interpretation of water quality data cannot provide meaningful conclusions unless based on the temporal and spatial variability of the hydrological regime.

Physical and Chemical Properties

Each freshwater body has an individual pattern of physical and chemical characteristics which are determined largely by the climatic, geomorphological and geochemical conditions prevailing in the drainage basin and the underlying aquifer. If surface waters were totally unaffected by human activities, up to 90-99 per cent of global freshwaters, depending on the variable of interest, would have natural chemical concentrations suitable for aquatic life and most human uses. Rare chemical conditions in freshwaters, such as occur in salt lakes, hydrothermal waters, acid volcanic lakes, peat bogs, etc., usually make the water unsuitable for human use In many regions groundwater concentrations of total dissolved salts, fluoride, arsenic, etc., may also naturally exceed maximum allowable concentrations.

Biological Characteristics

The development of biota (flora and fauna) in surface waters is governed by a variety of environmental conditions which determine the selection of species as well as the physiological performance of individual organisms. The primary production of organic matter, in the form of phytoplankton and macrophytes, is most intensive in lakes and reservoirs and usually more limited in rivers. The degradation of organic substances and the associated bacterial production can be a long-term process which can be important in groundwater and deep lake waters that are not directly exposed to sunlight.

2.2.   SURFACE WATER QUALITY AND POLLUTION ISSUES IN WATERSHEDS

A watershed is simply the geographical area that water flows across or through on its way to a common stream, river, or lake. A watershed can be very large (e.g. draining thousands of square kilometers to a major river or lake or the ocean), or very small, such as a hundred square hectares watershed that drains to a pond. A small watershed that lies inside of a larger watershed is sometimes referred to as a sub-watershed. Therefore a watershed can comprise different types of water bodies including rivers, streams, drainage systems, ponds, lakes and oceans. The water resources in a watershed can be divided into surface and ground water. Watersheds are characterized by interrelated water bodies and interrelated natural processes affecting the existing water bodies. It is not only natural processes that play a role in shaping the characteristics of watersheds: human interventions and activities also play an important role in altering these characteristics. The varying types of land use and anthropogenic activities influence both the surface and ground water quality in watersheds.

Water Quality and Pollution: Associated Definitions

One of the major problems within the water quality management field is a lack of a common understanding, or sometimes confusion in the use of water quality-related terminology which is related to the regulatory requirements and appropriate evaluation of water quality. Therefore it is important to understand and agree definitions and terms when dealing with water quality management. In view of the complexity of factors determining water quality, and the large choice of variables used to describe the status of water bodies in quantitative terms, it is difficult to provide a simple definition of water quality (Chapman, 1996).

A very general and brief definition could be: *Water quality* is a term used to describe the chemical, physical, and biological characteristics of water, usually in respect to its beneficial use and suitability for a particular purpose. So in assessing water quality, it is not just a list of chemical constituents to be investigated but the assessment should be linked to the water uses and there should be a targeted objective behind the assessment. Water quality assessment includes the use of monitoring to define the condition of the water, to provide the basis for detecting trends and to provide the information enabling the establishment of cause-effect relationships. Thus there is a logical sequence consisting of three components: monitoring, followed by assessment, followed by management.

Another important term to define is *pollution*. This can be defined briefly as an impairment of the beneficial use(s) of a water body. Finding chemical constituents in high concentrations in the water column or sediments is not pollution unless these constituents are impairing the beneficial uses of the water body. Therefore it is important to identify the uses of water body in order to comply with the permissible limits on the pollutants in it. However, the concept of pollution is relative, in that it reflects a change from some reference value to a value that causes problems for human use. A worldwide reference value is difficult to establish because insufficient monitoring has occurred prior to changes in water quality due to human activities. Furthermore, there is no universal reference of natural water quality because of the high variability in the chemical quality of natural waters (Meybeck *et al.*, 1996). Before we seek to understand the water quality or pollution of a certain water body as explained in these definitions, it is important to understand the following issues:

- The water or hydrologic cycle;
- The human interventions within the hydrologic cycle;
- The effect of these interventions on the hydrologic paths, and the water quality based on different uses.

The following section gives a brief explanation of these issues.

Water Cycle and Impacts of Anthropogenic Activities on Water Quality

The continuing growth in the global population is increasing the demand for freshwater. One important factor affecting freshwater availability is socioeconomic development, and another factor is the general lack of sanitation and waste treatment facilities in highly-populated and poor areas of developing countries. A principal cause of water scarcity is water quality degradation, which can critically reduce the amount of freshwater available for potable, agricultural, and industrial use, particularly in semi-arid and arid regions. Thus, the quantity of available freshwater is closely linked to the quality of the water, which may limit its use. Human activities and interventions related to the water resources within a watershed have a direct impact on altering the characteristics and the quality of the existing water bodies.

The quality of freshwater at any point in the watershed reflects the combined effects of many processes along water pathways. Human activities on all spatial scales affect both water quality and quantity. Alteration of the land use and associated vegetation not only changes the water balance, but typically alters the processes that control water quality. The effects of human activities on a small scale are relevant to an entire drainage basin. Furthermore, local, regional, and global differences in climate and water flow are considerable, implying various effects due to human activities on the land and on water quality and quantity, depending on the location within a watershed, geology, biology, physiographic characteristics, and climate. These natural characteristics also greatly control human activities, which will, in turn, modify (or affect) the natural composition of the water (Peters *and Meybeck.*, 2000).

The chemical composition of water varies depending on the nature of the solids, liquids, and gases that are either generated internally (in situ) or with which the water interacts. Furthermore, the chemical composition depends on the type of interaction. At the mostly pristine part of the hydrologic cycle, precipitation quality is derived from interactions with gases, aerosols, and particles in the atmosphere. Evaporation purifies water as vapor but concentrates the chemical content of the water from which it evaporated. Condensation begins the process of imparting chemical quality to atmospheric moisture by the inclusion of chemical substances through the dissolution of condensation nuclei. The complexity of the water-material interaction increases as precipitation falls on the land. The physical characteristics and mineralogical composition of soil and bedrock, topography, and biology substantially affect water quality. Most freshwater is a mixture of water derived from several hydrologic pathways. Figure (2-1) shows the hydrologic cycle and its components.

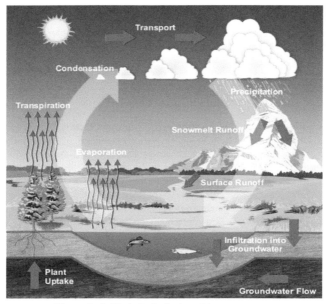

Figure (2-1): Hydrologic Cycle and its components

Furthermore, the stream water composition may change in situ due to biological reactions or due to the interactions with the streambed and adjacent riparian zone. Even the groundwater component of stream water is a mixture of water derived from different hydrologic pathways that vary in their composition due to the residence time of the water, the length of the hydrologic pathway, biological reactions, and the nature of the materials with which the water interacts. Temperature is another important variable that affects physical characteristics (e.g., transfer of gases), state changes (vapor, water, and ice), and chemical and biological reaction rates of the water. Natural water quality varies markedly and is affected by the geology, biology, and hydro-climatic characteristics of an area (Hem, 1985).

The quality of water changes as it moves through the hydrologic cycle from the sea, lakes, and rivers to its hydrosphere and then back to earth as precipitation, where the water again percolates through the soil or falls on the ocean, rivers, or lakes. (Changes that occur are physical, chemical, and biological.) These changes are either natural changes or changes due to human impacts and interventions. It is important to note for understanding how contamination of groundwater and surface water occurs, that it is necessary to understand the hydrologic cycle and the circulation of water from the ocean, atmosphere and land.

Human influences have had a direct effect on the hydrologic cycle by altering the land in ways that change its physical, chemical, and biological characteristics (Hem, 1985; Meybeck and Helmer, 1989). Physical alterations such as urbanization, transportation, farming (irrigation), deforestation and forestation, land drainage, channelization and damming, and mining alter hydrologic pathways and may change the water quality characteristics by modifying the materials with which the water interacts. In addition, these human activities alter water quality not only by changing hydrologic pathways, but by the addition of substances and wastes to the land. These activities include application of pesticides, herbicides, and fertilizers, and leaching to groundwater and surface water from landfills, mine tailings, and irrigated agricultural lands.

The chemical alteration associated with human activity is, in part, related to the physical alteration, but occurs mainly through the addition of wastes (gases, liquids, and solids) and other substances to the land. These additions include waste disposal on the land or in waterways and the application of substances to control the environment, such as fertilizers for crop production, herbicides for weed control, and pesticides for disease control. Atmospheric transport and deposition is a major hydrologic pathway of substances directly to surface water or indirectly to groundwater by infiltration through the soil. Human requirements for water also directly affect hydrologic pathways by providing water of a specified quality for different activities to sustain human existence (e.g., agriculture, potable supplies, power generation, power plant cooling, and industry).

Human activity is considered one of the most important factors affecting hydrology and water quality. Humans use large amounts of resources to sustain various standards of living, although measures of sustainability are highly variable depending on how sustainability is defined (Moldan *et al.,* 1997). Irrigated agriculture alone is responsible for about 75 percent of the total water withdrawn from "surface water and groundwater sources," and more than 90 percent of this water is consumed and delivered to the atmosphere by evaporation (United Nations Commission for Sustainable Development, 1997). In addition to placing a demand on the quantity of water, which is diverted for food production, the quality of water flowing through a typical agricultural area is markedly degraded. Degradation depends on several factors including the climatic characteristics and the various fertilizers and agrochemicals applied to increase yields. The excessive use of these fertilizers and agrochemicals has long-term effects on ground and surface water resources.

Each water use, including abstraction of water and discharge of wastes, leads to specific, and generally rather predictable, impacts on the quality of the aquatic environment. In addition to these intentional water uses, there are several human activities which have indirect and undesirable effects on the aquatic environment. Examples are uncontrolled land use for urbanisation or deforestation, accidental (or unauthorised) release of chemical substances and discharge of untreated wastes or leaching of toxic liquids from solid waste deposits into water bodies. It is also important to mention the structural interventions in the natural hydrological cycle through canalisation or damming of rivers, diversion of water within or among drainage basins. These interventions are usually undertaken with a beneficial objective in mind but their long term impacts on the water quality is obvious through experiences and cases all over the world.

Pollution Sources and Environmental Risks

Water quality reflects the composition of water as affected by nature and human cultural activities and uses, expressed in terms of both measurable quantities and descriptive statements. The quality status of receiving water bodies and their pollutants could be understood in a more comprehensive manner, expressed as 'integrity'. Integrity of the water body in this sense means taking into consideration all the affecting factors and parameters that directly impact the water quality or increase the pollution risk. Figure (2-2) shows the surface water body and all the principle factors and their component parameters that form its integrity.

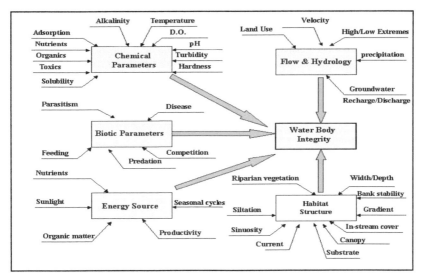

Figure (2-2): Five principal factors and their components that comprise the integrity of surface waters. (Adapted from Karr et al, 1986)

Sources or causes of water pollution can be classified into two types:

- Human alteration of the status of the water body and its inhabitants that downgrades its integrity and creates pollution
- Addition of allochthonous pollutant loads to the water body originating from outside of it.

The first cause of pollution may include a hydraulic modification of water bodies (channel lining and straightening that downgrade habitat, building dams and impoundments, flow diversion from streams, drainage of riparian wetlands, urban development that changes stream hydrology and cause stream bank erosion, in situ sediment contamination by human activities). The second cause, allochthonous sources of pollution (discharged from outside the water body), are identified as point and nonpoint sources.

The term **point source** means any discernible, confined and discrete conveyance, including but not limited to any pipe, ditch, channel, tunnel, conduit, well, discrete fissure, container, rolling stock, concentrated animal feeding operation, or vessel or other floating craft, from which pollutants are or may be discharged. This term does not include agricultural storm water discharges and return flows from irrigated agriculture. Major point sources under this definition include sewered municipal and industrial wastewater sources and effluents from solid waste disposal sites.

Non-point source water pollution arises from a broad group of human activities for which the pollutants have no obvious point of entry into receiving watercourses. They include everything else rather than point sources and include diffuse, difficult to identify, intermittent sources of pollutants usually associated with land use. A major example of nonpoint sources is the agricultural wastewater runoff and infiltration. In contrast, point source pollution represents those activities where wastewater is routed directly into receiving water bodies by, for example, discharge pipes, where they can be easily measured and controlled. Obviously, non-point source pollution is much more difficult to identify, measure and control than point sources. The non-point sources pollutants ultimately find their way into groundwater, wetlands, rivers and lakes and, finally, to oceans in the form of sediment and chemical loads carried by rivers. The ecological impact of these pollutants ranges from simple nuisance substances to severe ecological impacts involving fish, birds and mammals, and on human health.

The main characteristics of non-point sources include: they originate over a broad area (this makes them difficult to identify or to assess), they respond to hydrological conditions, are not easily measured or controlled directly (and therefore are difficult to regulate), and they are very much dependant on land and related management practices. Agriculture is only one of a variety of causes of non-point sources of pollution; however it is generally regarded as the largest contributor of pollutants of all the categories. Figure (2-3) illustrates the range and relative complexity of agricultural non-point source pollution is illustrated in. Table (2-1) outlines generally the classes of non-point sources and their relative contributions to pollution loadings.

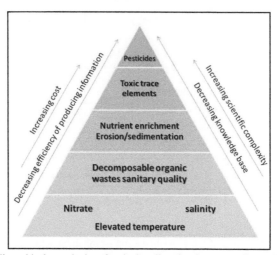

Figure (2-3): Hierarchical complexity of agriculturally-related water quality problems (Rickert, 1993)

Table (2-1): Classes of non-point sources and their associated pollutants

Non-point source	Definition	Associated pollutants
Agriculture Animal feedlots Irrigation Cultivation Pastures Dairy farming Orchards Aquaculture	Runoff from all categories of agriculture leading to surface and groundwater pollution. In northern climates, runoff from frozen ground is a major problem, especially where manure is spread during the winter. Vegetable handling, especially washing in polluted surface waters in many developing countries, leads to contamination of food supplies. Growth of aquaculture is becoming a major polluting activity in many countries. Irrigation return flows carry salts, nutrients and pesticides. Tile drainage rapidly carries leachates such as nitrogen to surface waters.	Phosphorus, nitrogen, metals, pathogens, sediment, pesticides, salt, BOD1, trace elements (e.g. selenium).
Forestry	Increased runoff from disturbed land. Most damaging is forest clearing for urbanization.	Sediment, pesticides.
Liquid waste disposal	Disposal of liquid wastes from municipal wastewater effluents, sewage sludge, industrial effluents and sludges, wastewater from home septic systems; especially disposal on agricultural land, and legal or illegal dumping in watercourses.	Pathogens, metals, organic compounds.
Urban areas Residential Commercial Industrial	Urban runoff from roofs, streets, parking lots, etc. leading to overloading of sewage plants from combined sewers, or polluted runoff routed directly to receiving waters; local industries and businesses may discharge wastes to street gutters and storm drains; street cleaning; road salting contributes to surface and groundwater pollution.	Fertilizers, greases and oils, faecal matter and pathogens, organic contaminants (e.g. PAHs Polycyclic Aromatic Hydrocarbons and PCBs Polycyclic CHL-aorinated Bi-

		Phenyls), heavy metals, pesticides, nutrients, sediment, salts, BOD, COD, etc.
Rural sewage systems	Overloading and malfunction of septic systems leading to surface runoff and/or direct infiltration to groundwater.	Phosphorus, nitrogen, pathogens (faecal matter).
Transportation	Roads, railways, pipelines, hydro-electric corridors, etc.	Nutrients, sediment, metals, organic contaminants, pesticides (especially herbicides).
Mineral extraction	Runoff from mines and mine wastes, quarries, well sites.	Sediment, acids, metals, oils, organic contaminants, salts (brine).
Recreational land use	Large variety of recreational land uses, including ski resorts, boating and marinas, campgrounds, parks; waste and "grey" water from recreational boats is a major pollutant, especially in small lakes and rivers. Hunting (lead pollution in waterfowl).	Nutrients, pesticides, sediment, pathogens, heavy metals
Solid waste disposal	Contamination of surface and groundwater by leachates and gases. Hazardous wastes may be disposed of through underground disposal.	Nutrients, metals, pathogens, organic contaminants.
Dredging	Dispersion of contaminated sediments, leakage from containment areas.	Metals, organic contaminants.
Deep well disposal	Contamination of groundwater by deep well injection of liquid wastes, especially oilfield brines and liquid industrial wastes.	Salts, heavy metals, organic contaminants.
Atmospheric deposition	Long-range transport of atmospheric pollutants (LRTAP) and deposition of land and water surfaces. Regarded as a significant source of pesticides (from agriculture, etc.), nutrients, metals, etc., especially in pristine environments.	Nutrients, metals, organic contaminants

Effect of Non-Point Sources of Pollution on Surface Water

As previously mentioned Irrigated agriculture is a significant and major source of surface water non point sources of pollution (NPS). The following are mainly the potential sources of non-point sources associated with irrigated agriculture:
• sediment
• nutrients
• pesticides
• salinity
• trace elements
• pathogens
• temperature (surface water quality impacts only)

Activities that can cause NPS pollution from irrigated agriculture include: new land development, cultural practices for production, pest management strategies, and irrigation practices. The main sources of NPS associated with irrigated agriculture and their effects on surface water can be summarized in the following items;

Sediment
Soil erosion and sediment deposition are primary causes of adverse impacts to surface water quality. Erosion is a natural process that can be accelerated by human activities. Slopes erode naturally, especially when vegetation is artificially removed. Sediment deposition occurs when the amount of sediment (solid material that has been transported from its site of origin by air, water, or gravity) exceeds the carrying capacity of the force that is moving it. Farmlands generally become a non-point source of pollution when farm operations remove a substantial amount of the vegetative cover, exposing the soil surface to the erosive action of water and wind. Eroded soil subsequently becomes sediment, creating the potential for water degradation.

Nutrients

The leaching of nutrients from watersheds into streams, lakes, and groundwater is a natural part of the nutrient cycle. When growers manipulate the soil-water-plant system to increase agricultural production, they can change the natural balance of the nutrient cycle. Nutrient sources associated with agricultural production practices include fertilizers, biodegradation of crop residues, agricultural and municipal wastes applied to land, and waste generated directly by animals. Nutrients from these sources become pollutants when they are transported off-site into nearby streams and lakes or when they percolate in excessive amounts to groundwater. Nutrients, whether dissolved or attached to soil particles, are transported by water. Soluble forms of nutrients leave their source sites by dissolving in water and travelling in solution with the runoff water or percolating soil water. Other forms of the nutrients that are attached to the soil, such as organic material, must be detached by erosion (surface water pollution only) before they can be transported. The two most significant nutrients affecting water quality are nitrogen (in the form of nitrate) and phosphorus (as phosphate). Surface water bodies contribute to a nutrient-rich environment, a condition called *eutrophication*. This process of increasing nutrients leads to increases in aquatic plants and algal blooms, which in turn depletes dissolved oxygen and so affects aquatic organisms.

Pesticides

Pesticides (insecticides, herbicides, fungicides) that move from their site of application into surface or groundwater can affect the usefulness of water through their potential to impact organisms other than their primary targets. The presence and bio-availability of pesticides in soil can adversely impact human and animal health, beneficial plants and soil organisms, and aquatic vegetation and animals. For example, herbicides can damage and destroy vegetation when they enter the aquatic system. Since this vegetation is cover and food for aquatic organisms, herbicides can affect an entire community. Dissolved oxygen levels that support aquatic life are sometimes reduced because of decaying plants killed by the herbicides. Ideally, a pesticide will stay in the treated area long enough to produce the desired effect and will then degrade into harmless materials. Four major processes affect the environmental fate of pesticides and therefore the risk of NPS pollution:

• *Retention*, which is the ability of the pesticide to stick (adsorb) to soil particles rather than leach away with soil water
• The *leaching rate* of water in the soil
• The ability of the pesticide to *degrade* over time
• The pesticide's potential for *volatilization* into the atmosphere

There are three main modes of degradation in soils: biological degradation (by means of microorganisms), chemical breakdown (by means of chemical reactions such as hydrolysis and redox reactions), and photochemical breakdown (by means of ultraviolet or visible light).

Pesticide retention in the soil is its ability to hold a pesticide in place and prevent it from being transported off site or to groundwater. *Adsorption*, the accumulation and adhesion of a pesticide onto the surface of soil particles is the primary means by which a soil retains a pesticide. Pesticide adsorption to soil particles depends on the chemical properties of the pesticide (i.e., its water solubility and polarity) and the soil properties (i.e., organic matter and clay content, pH surface charge characteristics, and permeability). For most pesticides, organic matter is the most important soil constituent determining the degree of adsorption. Pesticides, like nutrients, may be transported by water depending on their retention properties. Wind erosion or drift from pesticide applications may also contribute to the movement of pesticides away from their target area. Pesticides may enter surface waters in irrigation return flows and tile drainage either dissolved in the water or adsorbed to waterborne sediments.

Salinity

Irrigation water naturally contains a certain amount of dissolved minerals (salts). The amount of salt in the water depends on its source when irrigation water, regardless of its salt content, is applied to crops, the salts accumulate in the soil while the applied irrigation water is consumed by plants or lost to evaporation. To maintain the productivity of irrigated lands, accumulated salts must be leached below the root zone. For this reason, irrigation is always scheduled to exceed the anticipated crop consumption.

Trace Elements

Unlike nutrients and salts, trace elements typically do not originate from agricultural chemical applications. Rather, irrigation mobilizes naturally present trace elements. Trace elements are found at very low concentrations in all waters. Many trace elements are *at a low concentration* essential for human, animal, and plant health. At higher concentrations, however, they may become toxic to organisms. Trace elements may be mobilized in high concentrations along with other salts from marine sediments or soils with a naturally high salt content. They dissolve in percolating drainage water or groundwater and are discharged into wells or drainages, or in seepage into streams and lakes. The principal impact of trace element pollution is its adverse impact on animal life through biomagnification (the concentration of an element as it moves up the food chain) and on the degradation of water quality when used for human consumption. Trace elements of concern to irrigated agriculture include selenium, molybdenum, arsenic, vanadium, and boron. The water quality control requirements for trace elements are stringent and water quality objectives are usually prescribed in parts per billion. Because of its ability to impact beneficial uses of water at very low levels, the control of trace elements is complex and difficult. Trace element pollution must be examined on a case-by-case basis.

Pathogens

Pathogens are microorganisms and parasites that can cause illness in humans and in animals. A small subset of all pathogens, the *zoonotic pathogens,* are shed in the faeces of livestock and many wildlife species and can infect other animals as well as humans. These include *Salmonella, Giardia,* and *Cryptosporidium parvum,* and are the pathogens that cause concern with regard to food safety and water quality. The potential for pollution of surface waters increases when flows resulting from irrigation or rainfall come from land that has received untreated human or animal waste or when irrigation water contains animal manure. Localized contamination of surface water, groundwater, and the soil itself can result from animals in feedlots, corrals, exercise yards, pastures, and rangelands.

Other non-point sources of pathogens include septic tanks. The extent and concentration of pathogens in surface water depends largely on livestock density, timing and frequency of grazing, and access to streams and lakes. Pollution can occur when the daily rate of faecal deposition exceeds the ability of vegetated buffers, soil, and solar radiation (sunshine) to either filter out or inactivate the pathogenic microorganisms. Fecal coliform levels tend to increase as the intensity of livestock use increases. Maintaining the health of livestock is critical, as is proper management of the herd, its byproducts, and exposed land areas.

Temperature

Thermal pollution of surface waters has three basic sources that relate to irrig\ated agriculture:
- Development and subsequent cultural operations in irrigated agriculture can result in the loss of streamside vegetation that shades streams and helps to maintain the cool water temperatures required by many cold water fishes, especially trout, salmon, and steelhead.
- Drainage of irrigation water that has warmed while crossing a farm field can raise the temperature of a cold-water stream.
- Stream water diversions for irrigation and wetland management can lower overall stream flows.

Water Quality Main Parameters

Water quality is generally described according to biological, chemical and physical properties. Physical variables of water pollution include; color, temperature odor, total dissolved solids, suspended solids and turbidity. Biological pollutants include; bacteria (Faecal coliforms Total coliforms), viruses, parasites, algae. Chemical water quality variables include; nitrogen (e.g. nitrate, nitrite, ammonia), phosphorus dissolved oxygen (DO), pH (acidity, alkalinity), major ions (e.g. Ca^{++}, Na^+, Cl^-), pesticides (e.g. herbicides, insecticides, fungicides), heavy metals and toxic constituents, biochemical oxygen demand (BOD). In the practice of surface water resources engineering and water quality modeling, the following concentrations or variables are of importance according to Rafailidis (1994).

The Carbonaceous Biochemical Oxygen Demand (BOD). This is an indicator of the overall loading of the aquatic system due to the oxidation needs of organic pollutants. It also includes the respiration demand of aquatic microbes, which metabolize organic and fix inorganic matter (e.g. nitrates, inorganic phosphates, etc.).

The Dissolved Oxygen content (DO). Dissolved oxygen is a very important water quality parameter because it is essential for the respiration of the aquatic life. The maximum amount of oxygen that can be dissolved in water is called the oxygen solubility or dissolved oxygen saturation concentration (DOS), which decreases with an increase in temperature. Dissolved oxygen is utilised by dumped wastes in chemical oxidation reactions, or in biochemical reactions where organic material is biologically decomposed by aquatic microorganisms.This parameter is more critical because it shows whether there is sufficient oxygen in the water for aquatic life to survive. The actual DO content reflects the equilibrium between re-aeration at the surface added to photosynthetic oxygen generation by CHL-aorophyll in the water body, minus the biological and any chemical oxygen demand.

The concentrations of nutrients (ammonia, nitrates, phosphates, inorganic nitrogen or phosphorous), these are linked directly to non-point sources of agricultural watersheds, as a consequence of soil fertilization, insecticides or pesticide use, etc. Aquatic microorganisms metabolize nutrients and the inorganic elements are fixed to more complex compounds. Algae play a very important role in these processes, enhancing water denitrification (release of N_2 to the atmosphere) or nitrification (capture of N_2 from air). On the other hand, simultaneous presence of phosphorous enhances algal growth, leading to eutrophication, i.e., abnormal growth of algae and aquatic flora. This is particularly troublesome in enclosed waters (e.g., lakes, lagoons) but also occurs in coastal areas suffering from large pollution inflows and suppressed natural circulation flushing.

The coliform bacteria concentration. Although these micro-organisms are not pathogenic and exist naturally in the intestine of humans, their presence indicates pollution due to urban sewage effluents. Upon discharge into the water body environmental conditions such as temperature and sunlight determine the eventual fate of coliform bacteria through a multitude of processes (e.g., photo-oxidation, sedimentation, ph, algae, etc.).

Apart from the above pollutants, sediments in the water column may also cause environmental problems as they bury benthic flora and choke the gills of aquatic animals. Therefore, it is concluded that BOD in surface waters indicates the overall organic pollution of the water, and DO shows whether the aquatic life may be sustained there, whereas nutrient concentration gives the potential for eutrophication. Coliform counts indicate the danger of disease for humans using the water for any purpose of use.

Other very important water quality indicators are the *flora and fauna*. The growth of aquatic plants, both rooted and phytoplankton (mainly algae) is controlled by light, nutrients and animal grazing. Light enables photosynthesis, which produces oxygen. The most important nutrients as mentioned earlier are *nitrogen, carbon* and *phosphorous.* Nitrogen and phosphorous are both considered most important pollutants in non-point sources in agricultural watersheds. Therefore, the coming section explains in details the nitrogen and phosphorous cycles.

Nitrogen Cycle

Brown and Jhonson (1991) give details of the effect of nitrogen on water quality. The nitrogen cycle is a representation of the various forms of N and how they relate to one another through many complex interactions. Figure (2-4), a simplified version of the nitrogen cycle, illustrates many of the complex interactions of various forms of nitrogen, including:

atmospheric nitrogen (N^2), ammonia (NH$_3$), ammonium ion (NH$_4^+$), nitrite ion (NO$_2^-$), and nitrate ion (NO$_3^-$). Each nitrogen form has characteristics that relate to plant utilization and possible impacts on water resources.

Nitrogen Availability to Plants

In their need for nitrogen, non-leguminous plants, such as lawn and turf grasses, corn and most fruit and vegetable crops, must rely on either bacteria that live in the soil to "fix" the nitrogen (N^2) into a usable form or nitrogen from decomposing organic matter, or fertilizers. The forms of nitrogen that most plants can use are ammonium ion (NH$_4^+$) and nitrate ion (NO$_3^-$), as shown in Figure (2-4). Of these, the ammonium and nitrate ions are the most common forms taken in through plant roots.

Ammonium is converted to the nitrite and nitrate forms rather quickly by nitrifying bacteria, such as *Nitrosomonas .sp* and *Nitrobacter .sp*, which add oxygen to the ammonium ion and convert it to nitrate. However, the legumes, for example, alfalfa, clover, soybeans and peanuts, have nodules on their roots that contain bacteria. The plants benefit by having the bacteria that fix atmospheric nitrogen into a usable form for the plant, while the bacteria benefit from the energy obtained in the chemical conversion. It is to be taken into consideration that the ammonia and nitrite forms of nitrogen are highly toxic to humans.

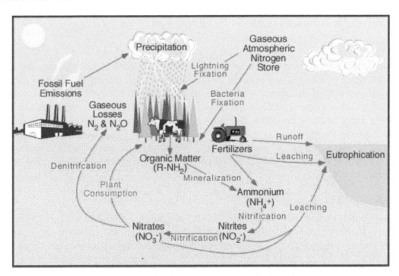

Figure (2-4): The nitrogen cycle in soil

Nitrogen Loss from Availability to Plants

Nitrogen can become unavailable to plants primarily in three ways. First, most nitrogen is lost through denitrification, which is a problem in wet or compact soils. Since these soils contain little oxygen, denitrifying bacteria remove the oxygen from nitrite (NO$_2^-$) and nitrate (NO$_3^-$) ions for their own use, releasing N^2 and/or N^{2}O back to the atmosphere. The second means of nitrogen loss is by nitrate leaching, which is a particular concern with the nitrate ion (NO$_3^-$). Leaching occurs when the water-soluble nitrate ion moves through the soil as water percolates downward beyond the reach of plant roots. Surface volatilization (conversion to the gaseous phase) is the third method of nitrogen loss. This loss occurs when ammonia (NH$_3$), usually in the form of urea, volatilizes and is lost to the atmosphere. Surface volatilization is usually a problem in areas with high temperatures, and with soils that have a high pH value. Soils that have been compacted by field operations and other human activities also are a problem because it may not be possible to properly mix the urea with the compacted soil. Another pathway for nitrogen loss from plant availability is the loss of the nitrogen through the process of soil erosion by water.

Nitrogen Cycle-Hydrologic Cycle: Interactions

Since nitrogen and water are so vital for all organisms, it is inevitable that components of the nitrogen and hydrologic cycles are closely related. These relations have particular importance for agriculture, and lawn and turf management. By understanding these interactions, we can better understand the effects of human activities on water resource quality. Nitrogen, mostly in the form of ammonium and nitrate, reaches the Earth's surface as a result of atmospheric lightning, precipitation and industrial pollution, this is accounted for as atmospheric production.

Denitrification

Nitrifying organisms can only function when free oxygen (O_2) is present. In saturated soils, free oxygen is very low, suppressing the growth of the nitrifying organisms, often causing nitrogen deficiencies in excessively wet soils. This condition is enhanced by denitrifying bacteria since they thrive in an oxygen-free environment, like a saturated soil, and therefore consume nitrate at a rapid rate. Excessive rainfall promotes nitrogen loss not only by promoting nitrate leaching from the plant root zone, but also by creating wet soil conditions that favor denitrification. Evaporation works in the opposite way to remove water from the upper soil layers. Space then becomes available for oxygen, thereby making the environment suitable for the growth of nitrifying bacteria.

Surface Volatilization

In agricultural situations, surface volatilization (vaporization of urea to ammonia gas) may occur when urea is applied on crop residues, and is not in good contact with soil particles. To limit volatilization of the urea, producers usually incorporate it into the soil by tillage to bring the urea into contact with the soil. Limited rainfall also helps with proper incorporation of the urea in the upper portion of the soil profile. When water and urea combine, the result is the ammonium ion (NH_4^+), which has a positive charge and attaches to negatively charged soil particles. Both tillage and rainfall can help make nitrogen available for plant use. Unfortunately, the interaction between tillage and excessive rainfall increases the potential for soil erosion. After tillage, the soil is more susceptible to being carried away by water during heavy rainfall.

Nitrogen Movement through the Soil

The nitrate ion (NO_3^-) is the most water-soluble form of nitrogen as well as the form least attracted to soil particles. Therefore, its interaction with the hydrologic cycle is very important since it moves where water moves. Precipitation, evaporation and transpiration may affect the movement of nitrate in the near-surface soil profile. Rainfall that infiltrates the soil surface may cause nitrate ions to move down through the soil profile by percolation. The more rain that infiltrates into the soil, the further down in the profile nitrate ions move. Nitrate movement below the plant root zone is called nitrate leaching. Soil texture, structure and permeability, along with other soil properties, affect nitrate leaching. Deep percolation of water through the soil profile potentially allows the movement of nitrate out of the root zone and downward, where it may pollute the underlying aquifer. In contrast to the nitrate ion, the ammonium ion has a strong attraction for soil, and therefore is considered to be immobile in most soils. However, in soils with very high sand and low organic matter contents, the ammonium ion will move in the direction of water movement.

Surface evaporation and transpiration may help nitrate move toward the soil surface within the root zone as a result of capillary movement as the plant withdraws water from the soil profile. Upward movement of nitrate occurs mainly in the summer when evaporation and transpiration exceed rainfall. Once nitrates get into the surface water or the groundwater, the greatest concerns are for infants less than one year old and for young or pregnant animals. High levels of nitrates can be toxic to newborns, causing anoxia, or internal suffocation.

Nitrogen Movement to Surface Waters

Runoff contributes to the movement of several forms of nitrogen to surface water. Runoff results when the rainfall rate exceeds the infiltration rate at the soil surface. Runoff from agricultural and suburban watersheds carries sediment, as well as nutrients like nitrate and ammonium. Ammonium ions attach to sediments very readily, which means they move with soil, but generally do not leach. Therefore, ammonium may contribute to surface-water problems, but generally does not impact groundwater.

Subsurface drainage improvements may contribute to the movement of the nitrate form of nitrogen to surface waters. Many agricultural soils with poor internal drainage require the installation of drainage systems to promote a healthy environment for crop root development, and to improve nitrogen efficiency. Where nitrate is present in wet agricultural soils without proper drainage improvement, there is a great potential for nitrogen loss by denitrification if soil conditions (i.e., organic matter and temperature) are favorable. In addition, rapid removal of excess water from the plant root zone decreases the potential for denitrification.

With subsurface drainage, some of the rainfall that infiltrates the soil surface is intercepted by the subsurface drainage system, and subsequently discharged to a ditch or stream and consequently to lakes at the downstream end. If nitrate ions are present in the soil profile, they will move with the percolating water. Subsurface drainage systems actually intercept the nitrate after it has been leached from the plant root zone, and before it has the opportunity to move by deep percolation to an underlying aquifer. Unfortunately, these systems may discharge nitrate in surface waters instead. Subsurface drainage water generally will have a higher concentration of nitrate than runoff water, but considering the greater potential for movement of sediment, nitrate, ammonium and phosphorous in runoff, subsurface drainage water is generally of better quality. The loss of nitrate in subsurface drainage water is not a simple matter to resolve since it is related to rainfall timing and amount, soil profile characteristics, subsurface water flow rate (soil-dependent), nitrogen application rate and timing, and the extent of plant uptake of the nitrate available in the soil profile.

Nitrogen transformation processes are affected by the availability of oxygen and organic carbon plus the presence of microorganisms such as Nitrosomonas and Nitrobacter in the soil. These processes can go on simultaneously, coexisting in close proximity, and varying temporally in the same setting. It is the balance between these processes and their seasonal timing that determines how much nitrogen is available for crops and how much nitrogen may be lost from the soil to the atmosphere or surface and ground   water.

Phosphorous Cycle

The phosphorus cycle originates with the introduction of phosphate (PO_4) into soils from the weathering of rocks. Phosphate enters living ecosystems when plants take up phosphate ions from the soil.   Figure (2-5) shows the phosphorous cycle. Phosphate moves from plants to animals when herbivores eat plants, and when carnivores eat herbivores. The phosphate that has been taken up into the tissue of animals is returned to the soil through the excretion of urine and faeces, as well as through decomposition of dead animals. Phosphate in plants also returns to the soil through decomposition.

These same processes occur in aquatic systems as well. Plants take up waterborne phosphate, which then travels up through successive stages of the aquatic food chain. Phosphate that is not taken up into the food chain, as well as that in dead and decomposing organisms, settles on the ocean floor or on lake bottoms. When these sediments are stirred up, this phosphate may re-enter the biological phosphorus cycle, but much more of it is buried in the sedimentary rock. Like nitrogen, phosphorus is considered a pollutant when it occurs in excess concentrations in surface waters. Phosphorus can contribute to over-fertilizing, or eutrophication of these waters. Unlike nitrogen, phosphorus is not highly soluble. It binds tightly to molecules in the soil and so can build up to harmful levels. It reaches surface waters not by dissolving but by traveling with soil particles in runoff to which it has bonded. Phosphates were once commonly used in laundry detergents, which contributed to excessive concentrations in rivers, lakes, and streams. Most detergents no longer contain phosphorous. The predominant sources of phosphorous in bodies of water are agricultural and excessive fertilizers and improperly disposed animal wastes.

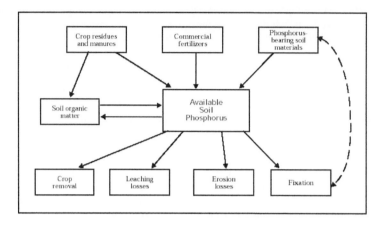

Figure (2-5): The phosphorous cycle

Human Inputs to the Phosphorus Cycle

Human influences on the phosphate cycle come mainly from the introduction and use of commercial synthetic fertilizers. The phosphate is obtained through mining of certain deposits of calcium phosphate called apatite. Huge quantities of sulfuric acid are used in the conversion of the phosphate rock into a fertilizer product called "super phosphate". Plants may not be able to utilize all of the phosphate fertilizer applied; as a consequence, much of it is lost from the land through the water run-off. The phosphate in the water is eventually precipitated as sediments at the bottom of the body of water. In certain lakes and ponds this may be redissolved and recycled as a problem nutrient. Animal wastes or manure may also be applied to the land as fertilizer. In certain areas very large feed lots of animals, may result in excessive run-off of phosphate and nitrate into streams. Other human sources of phosphate are in the outflows from municipal sewage treatment plants. Without an expensive tertiary treatment, the phosphate in sewage is not removed during various treatment operations. Again an extra amount of phosphate enters the water.

2.3.   IRRIGATED WATERSHEDS AND CATCHMENT–LAKE SYSTEMS: WATER QUALITY ISSUES

The physically complex watersheds in general may have some common features such as similar physical components of the watershed including the drainage catchment, downstream catchment connection to lakes, lagoons or direct connection to sea or ocean. The geographic location, hydrology, land use and associated pollution problems are considered the specific characteristics of a watershed. Freshwater systems in these watersheds are influenced by changing land-use patterns in the whole watershed area. The pattern and extent of cities, roads, agricultural land, and natural areas within a watershed influences infiltration properties, transpiration rates, and runoff patterns, which in turn impact water quantity and quality. For example, expanding agricultural land increases the load of nutrients and fertilizers from crops that end up in rivers and streams, impacting the water quality and the biodiversity of freshwater ecosystems. In addition, because most of the fertile lands are already under agriculture use, the areas for agriculture expansion are areas that require irrigation and usually other inputs such as fertilizers. In many cases these agricultural practices cause soil erosion, water logging, salinization, and water pollution from runoff that reduce the capacity of the soil to produce crops in a sustainable way and also damage freshwater ecosystems.

Considering irrigated watersheds, where the main land use is agricultural activities, historically, attention to water quality in irrigated watersheds has been focused on irrigation waters and the relationships of their chemical composition to soil permeability and crop production. Recently, because of environmental concerns, it has become necessary to look beyond the quality of irrigation waters and consider also the quality of waters that drain from irrigated lands. Irrigation agriculture affects drainage-water chemical composition. In turn, drainage waters can influence the quality of receiving waters which may have a variety of beneficial uses to be protected (Branson et al.1975).

The two types of drainage waters from irrigated lands, surface runoff and subsurface drainage or percolation water are characteristically different in composition and chemical concentration. The pollution potential of subsurface drainage waters, with respect to nitrate and total soluble salts is of particular concern. Studies of individual field conditions are providing information that can be extrapolated to show the effects of watershed management on ultimate water quality in a receiving stream lake or ground water.

Since the main focus of this research is on irrigated or agricultural watersheds including connected surface water systems, it is important to highlight the effect of agriculture on pollution of these systems. Agricultural production can generate contaminants that can have many negative effects on surface or ground water supplies. Contaminants that are associated with cropping and livestock practices include sediment, nutrients (nitrogen and phosphorus) from inorganic fertilizers and organic livestock wastes, crop protection chemicals such as herbicides and insecticides, microorganisms from livestock wastes, and salts and trace elements from irrigation residues. Contaminants are transported, either attached to sediment or dissolved in drainage water, to surface and ground water through all phases of the water or hydrologic cycle. Impaired water quality can restrict water uses for such activities as stock watering and irrigation, drinking water supplies, fisheries and other aquatic life. In summary, agricultural activities, in addition to other land uses, can have a significant impact on the quality and uses of our water resources.

Focusing on catchment-lake systems, the problem of pollution and impaired water quality becomes more sophisticated due to the accumulative pollution from the drainage systems that ends up into the connected lakes. Aquatic ecosystems are dynamic and complex natural systems. Erosion and transport processes primarily control change in flowing water systems, whereas internal recycling of inorganic and organic materials and inputs from the watershed primarily drive changes in lake systems. Therefore, it is crucial to understand the physical and hydrodynamic characteristics of the water bodies within the watershed. The following section focuses on the general features of the lakes and the specific features of shallow lakes as the downstream end of the catchment receiving all drainage water and contaminants.

2.4.   LAKES AND ENVIRONMENTAL RISKS

There is not a specific sharp definition of lakes but in the next section an attempt is made to put some characteristics and features to define lakes and distinguish lakes from other surface water bodies. A lake (from Latin *lacus*) is a terrain feature (or physical feature), a body of liquid on the surface of earth that is localized to the bottom of watershed. On Earth, a body of water is considered a lake when it is inland, not part of the ocean, is larger and deeper than a pond, and is fed by a river. Several attempts in literature are made to define lakes. According to (Thomas et al., 1996), a lake may be defined as an enclosed body of water (usually freshwater) totally surrounded by land and with no direct access to the sea. A lake may also be isolated, with no observable direct water input and, on occasions, no direct output. In many circumstances these isolated lakes are saline due to evaporation or groundwater inputs. Depending on its origin, a lake may occur anywhere within a river basin. A headwater lake has no single river input but is maintained by inflow from many small tributary streams, by direct surface rainfall and by groundwater inflow. Such lakes almost invariably have a single river output. Further downstream in river basins, lakes have a major input and one major output, with the water balance from input to output varying as a function of additional sources of water.

Lakes may occur in series, inter-connected by rivers, or as an expansion in water width along the course of a river. In some cases the distinction between a river and a lake may become vague and the only differences may relate to changes in the residence time of the water and to a change in water circulation within the system. In the downstream section of river basins, lakes (as noted above) are separated from the sea by the hydraulic gradient of the river, or estuarine system. The saline waters of the Dead Sea, the Caspian and Aral seas are, therefore, strictly lakes whereas the Black Sea, with a direct connection to the Mediterranean via the Sea of Marmara, is truly a sea.

Characteristics and Classification of Lakes

Lakes are traditionally under-valued resources to human society. They provide a multitude of uses and are prime regions for human settlement and habitation. Uses include drinking and municipal water supply; industrial and cooling water supply; power generation; navigation; commercial and recreational fisheries; body contact recreation, boating, and other aesthetic recreational uses. In addition, lake water is used for agricultural irrigation, and for waste disposal. Good water quality in lakes is essential for maintaining recreation and fisheries and for the provision of municipal drinking water in some areas of the world. These uses are clearly in conflict with the degradation of water induced by agricultural use and by industrial and municipal waste disposal practices. The management of lake water quality is usually directed to the resolution of these conflicts (Thomas *et al.*, 1996).

Origin and Types of Lakes
In geological terms lakes are ephemeral. They originate as a product of geological processes and terminate as a result of the loss of the ponding mechanism, by evaporation caused by changes in the hydrological balance, or by in filling caused by sedimentation. A detailed review on the origins of lakes and different types was done by (Hutchinson, 1957). In this review he differentiated 11 major lake types, sub-divided into 76 sub-types. Meybek, 1995, gave the following a summary on this review;

Glacial lakes: Lakes on or in ice, pounded by ice or occurring in ice-scraped rock basins. The latter origin (glacial scour lakes) contains the most lakes. Lakes formed by moraines of all types, and kettle lakes occurring in glacial drift also come under this category. Lakes of glacial origin are by far the most numerous, occurring in all mountain regions, in the sub-arctic regions and on Pleistocene surfaces. All of the cold temperate, and many warm temperate, lakes of the world fall in this category (e.g. in Canada, Russia, Scandinavia, Patagonia and New Zealand).

Tectonic lakes: Lakes formed by large scale crustal movements separating water bodies from the sea, e.g. the Aral and Caspian Seas. Lakes formed in rift valleys by earth faulting, folding or tilting, such as the African Rift lakes and Lake Baikal, Russia. Lakes in this category may be exceptionally old. For example, the present day Lake Baikal originated 25 million years ago.

Fluvial lakes: Lakes created by river meanders in flood plains such as oxbow and levee lakes, and lakes formed by fluvial damming due to sediment deposition by tributaries, e.g. delta lakes and meres.

Shoreline lakes: Lakes cut off from the sea by the creation of spits caused by sediment accretion due to long-shore sediment movement, such as for the *coastal lakes of Egypt.*

Dammed lakes: Lakes created behind rock slides, mud flows and screeds. These are lakes of short duration but are of considerable importance in mountainous regions.

Volcanic lakes: Lakes occurring in craters and calderas and which include dammed lakes resulting from volcanic activity. These are common in certain countries, such as Japan, Philippines, Indonesia, Cameroon and parts of Central America and Western Europe.

Solution lakes: Lakes occurring in cavities created by percolating water in water soluble rocks such as limestone, gypsum or rock salt. They are normally called Karst lakes and are very common in the appropriate geological terrain. They tend to be considered as small, although there is some evidence that some large water bodies may have originated in this way (e.g. Lake Ohrid, Yugoslavia). Excluding reservoirs, many other natural origins for lakes may be defined, ranging from lakes created by beaver dams to lakes in depressions created by meteorite impact.

Classification of Lakes

Lakes could be classified according to their origin as mentioned. Two other systems of classification which are based upon processes within lakes, and which are used universally, provide the basis upon which assessment strategies and interpretation are based. These are shown in the following section; the *physical or thermal lake classification* and the *classification by trophic level*.

Physical/thermal lake types

The uptake of heat from solar radiation by lake water, and the cooling by convection loss of heat, result in major physical or structural changes in the water column. The density of water changes markedly as a function of temperature, with the highest density in freshwater occurring at 4 °C. The highest density water mass usually occurs at the bottom of a lake and this may be overlain by colder (0-4 °C) or warmer (4-30 °C) waters present in the lake. A clear physical separation of the water masses of different density occurs and the lake is then described as being stratified. When surface waters cool or warm towards 4 °C, the density separation is either eliminated or reaches a level where wind can easily induce vertical circulation and mixing of the water masses producing a constant temperature throughout the water column. In this condition the lake is termed homothermal and the process is defined as vertical circulation, mixing, or overturns. The nomenclature applied to a stratified lake is summarized in three levels:

- *The Epilimnion*: or surface waters of constant temperature (usually warm) mixed throughout by wind and wave circulation,
- *The Hypolimnion:* the deeper high density water or (this is usually much colder, although in tropical lakes the temperature difference between surface and bottom water may be only 2-3 °C),
- *The Metalimnion*: a fairly sharp gradational zone between the two, normally referred to as thermocline.

The differentiation between shallow and deep lakes depends on the thickness of the epilimnion layer, this in run is dependent on the lake surface area, solar radiation, air temperature and lateral circulation and movement of the surface water. Commonly, it extends to about 10 m depth but in large lakes it can extend up to 30 m depth. Stratification in very shallow lakes is generally rare since they have warm water mixing throughout their water column due to wind energy input. However, winter or cold water stratification can occur even in the shallowest lakes under the right climatic conditions.

The interpretation of a shallow lake has never been satisfactorily defined, although there is a relationship between lake depth and surface area which controls the maximum depth to which wind induced mixing will occur. Therefore, an acceptable definition of a shallow lake is one which will overturn and mix throughout its water column when subjected to an average wind velocity of 20 km h⁻¹ for more than a six hour period. As a general rule, wind exposed lakes of 10 m depth or less are defined as shallow water lakes (Thomas *et al.*, 1996)

Trophic Status

The concept of trophic status as a system of lake classification was introduced by early limnologists such as Thienemann (1925, 1931), and has been subject to continuous development up until the present time (Vollenweider, 1968). The process of eutrophication underlying this scheme is one of the most significant processes affecting lake management and is, therefore, described in more detail. The underlying concept is related to the internal generation of organic matter which is also known as autotrophic production. External inputs of organic matter from the watershed (allotrophy) produce dystrophic lakes rich in humic materials. Such lakes may also be termed brown water or polyhumus lakes. In these lakes, most of the organic matter is derived from the surrounding watershed and internal carbon production is generally low. Detailed discussion on eutrpophication and its impact on lake water quality is given below.

The classification of lakes based on trophic level is shown in below. The names given to the classifications represent empirically defined intervals ranging from very low to very high productivity, which are defined as follows:

Oligotrophic lakes: Lakes of low primary productivity and low biomass associated with low concentrations of nutrients (N and P). In temperate regions the fish fauna is dominated by species such as lake trout and whitefish. These lakes tend to be saturated with oxygen throughout the water column.

Mesotrophic lakes: These lakes are less well defined than either oligotrophic or eutrophic lakes and are generally thought to be lakes in transition between the two conditions. In temperate regions the dominant fish may be whitefish and perch. Some depression in O_2 concentrations occurs in the hypolimnion during summer stratification.

Eutrophic lakes: Lakes which display high concentrations of nutrients and an associated high biomass production, usually with a low transparency. In temperate regions the fish communities are dominated by perch, roach and bream. Such lakes may also display many of the effects which begin to impair water use. Oxygen concentrations can get very low, often less than 1 mg l-1 in the hypolimnion during summer stratification.

Hypereutrophic lakes: Lakes at the extreme end of the eutrophic range with exceedingly high nutrient concentrations and associated biomass production. In temperate regions the fish communities are dominated by roach and bream. The use of the water is severely impaired as is described below. Anoxia or complete loss of oxygen often occurs in the hypolimnion during summer stratification.

Dystrophic lakes: As defined previously, these are organic rich lakes (humic and fulvic acids) with organic materials derived by external inputs from the watershed.

Water Quality and Pollution Issues in Lakes

A water quality issue within a lake may be defined as a water quality problem or impairment that adversely affects the lake water to an extent which inhibits or prevents some beneficial water use. Since a water quality issue normally results from the harmful effects of one or more human uses within the lake itself or within the lake's watershed, major conflicts between users or uses may occur in lake systems subjected to multiple uses. Therefore we can say that a lake is a reflection of its watershed. More specifically, a lake reflects the watershed's size, topography, geology, land use, soil fertility and erosion, and vegetation. The impact of the watershed is evident in the relation of nutrient loading to the watershed/lake surface area ratio. Figure (2-6) shows the relation between the watershed area and the lake area ratio and the nutrients loadings (Monson, 1992).

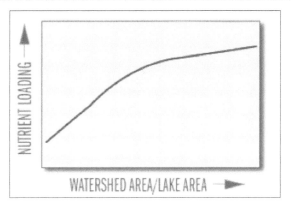

Figure (2-6): The relation between the watershed area and the lake area ratio and the nutrients loadings (After Monson, 1992)

A number of issues affect lake water quality. These have mostly been identified and described in industrialised regions, such as the North American Great Lakes, which have progressed from one issue to another in a sequence parallel to social and industrial development (Meybeck *et al.*, 1989). Meybeck *et al.* (1989) have highlighted that the developing nations of today are responsible for managing lakes which are being subjected to synchronous pollution from the simultaneous evolution of rural, urban and industrial development, differently from the historical progression which occurred in developed societies. This means that, in the developing world, multi-issue water quality problems must be faced with greater cost and complexity in assessment design, implementation and data interpretation. The current issues facing lake water quality can be summarized in the following:

- Eutrophication
- Health related issues and organic wastes
- Contaminants
- Lake acidification
- Salinisation

In this presented research work, the main focus is on eutrophication of shallow lakes. The next section describes characteristics of shallow lakes, the eutrophication as a water quality problem, its reasons and impacts and the processes involved.

2.5.   EUTROPHICATION OF SHALLOW LAKES

In order to define the eutrophication problems of shallow lakes, and how the problem originates from the upstream activities and land uses in the watersheds, there are some definitions characteristics that should be highlighted. First we have to define the ***Ecosystem***, which is the unit of natural organization in which all living organisms interact collectively with the physical chemical environment as one physical system. Lakes are considered living ecosystems. Lake ecosystems are complex, involving both terrestrial and aquatic photosynthesis, external and internal nutrients, grazer and detritus food webs, and aerobic and anaerobic metabolism. The lake ecosystem consists of two major components: the "aquatic" component which is the overall water body itself, and the "paralimnetic" component which consists of the drainage basin or watershed. The paralimnetic component could be divided into a variety of land-use. Likewise, the aquatic component could be divided into littoral zone, pelagic zone, benthic boundary layer and sediments. Specific components of the lake ecosystem are shown in figure (2-7) and could be distinguished as follows:

- The littoral zone is the near shore area where sunlight penetrates all the way to the sediment and allows aquatic plants (macrophytes) to grow.

- The limnetic zone is the open water area where light does not generally penetrate all the way to the bottom.
- The euphotic zone of the lake, which is the layer from the surface down to the depth where light levels become too low for photosynthesis energy producers.
- The benthic zone, which is the bottom sediment, it has a surface layer abundant with organisms.

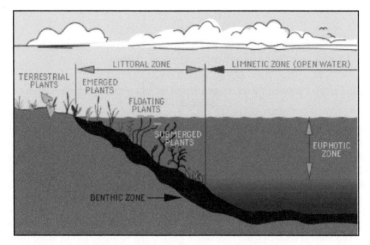

Figure (2-7): The typical lake distinct zones of biological communities
linked to the physical structure of the lake.

Shallow Lakes

In several parts of the world shallow lakes are more abundant than deep ones. Numerous shallow lakes, for instance, are found at the edge of the ice cover during the Weichselian glaciation period. Also, human activities such as digging for peat, sand, gravel, or clay have produced considerable numbers of shallow lakes and ponds. The term 'wetlands' is often used to refer to shallow lakes and adjacent marshy land. Such habitats are notoriously rich in wildlife. In densely populated areas even small lakes can be very important from a recreational point of view. Fishing, swimming, boating and bird watching attract a large public (Scheffer, 2001). In other areas the image is totally different where the lakes could be mainly used for commercial fishing, discharging drainage and waste water from catchments and fish ponds.

The pristine state of the shallowest lakes is probably one of clear water and rich aquatic vegetation. Nutrient loading has changed this situation in many cases. The lakes have shifted from clear to turbid, and with the increase in turbidity, submerged plants have largely disappeared. Efforts to restore the clear state by means of reduction of the nutrient loading often are unsuccessful. This has invoked experiments with additional methods such as temporary reduction of the fish stock. This has catalyzed the development of insights into the mechanisms that govern the dynamics of shallow lake communities (Scheffer, 2001).

Shallow lakes as defined earlier in this section are forced by wind mixing with depth of ten meters or less. They typically, fall into the eutrophic category, although some might also be mesotrophic. Shallow lakes can have the following general characteristics.

Vegetation:
Aquatic plant growth is abundant due to high nutrient content (phosphorous, nitrogen, and minerals) and the shallowness of the water. Aquatic plants need both nutrients and sunlight to grow. These

plants provide excellent food and habitat for zooplankton, insects, fish, waterfowl and other wildlife. They also lock up sediments, helping keep the water more clear.

Depth:
Shallow lakes (and deeper wetlands) are often less than five meters deep, although in some cases they might be as deep as ten meters. Most lakes do have some shallow lake components to them in bays.

Shoreline zone:
Technically referred to as the littoral zone, this zone is evident by the stands of emergent aquatic plants such as cattails, bulrush, and reeds that are present, as well as submerged plants such as coontail water lily, potamogeton species and floating plants such as water hyacinths. Healthy shallow lakes have plants growing throughout the entire basin, creating an extended littoral zone. Nutrient/sediment mixing: Sediment and nutrients in shallow lakes, unlike in deeper lakes, are constantly mixing. In a shallow lake where there is not much of a temperature difference and nutrients and sediments can easily be stirred up by wind, wave action and undesirable fish species.

Fluctuating water levels:
Shallow lakes can often benefit from periods of low water that stimulates beneficial aquatic plant growth. When water levels remain too high, too long, plant growth can be obstructed and water quality deteriorates. Algae growths usually result from such conditions.

Fish:
Low water conditions can help set the stage for winterkills that can decrease or eliminate populations of undesirable fish species such as carp and black bullhead. While shallow lakes can support populations of game fish, low levels of dissolved oxygen and winterkills tend to limit their numbers.

Land use impacts:
Agricultural chemicals as mentioned earlier, run-off from agricultural lands, drainage water and soil particles that flow into a shallow lake will eventually cause the lake to become seriously degraded. Shallow lakes can be more susceptible to such run-off than their deeper water lakes.

Surface water use:
Surface water use can sometimes be as important as land use management in maintaining a healthy shallow lake. Aquatic vegetation can suffer from too many fishing nets docks, boats and outboard motors on a lake. It's important to note that shallow lakes, even though they share common characteristics, are each unique in their own way. The surrounding terrain, land use practices, and lake use are some of the considerations to be taken into account when dealing with management of shallow lakes. Specific properties of shallow lakes and their surrounding catchments that should be defined thoroughly are the ***aqautic vegetation*** and ***commercial fisheries or fish ponds***. These components of the lake physical system have direct impact on the lake water quality and trophic status.

Aquatic Vegetation

The vegetation of water and swamp habitats creates a complex area in the ecotone zone between water and land, where the transport of allochtonic matter takes place (Joniak *et al.,* 2007). The nearshore or meadow vegetation may be a barrier restricting the distribution of anthropogenic contamination (e.g., Szpakowska, 1999). In relation to aquatic macrophytes, opinions considering the role of macrophytes in modyfying the physical-chemical features of their environment have as yet not been defined. On one hand the dynamics of water in shallow reservoirs or lakes is often determined by irregular processes e.g., floods, wind mixing etc., which have a great influence on water parameters (Joniak et al., 2000). On the other hand overgrowing macrophytes may intake nutrients from water or restore them from sediments into the water (Grane´ li & Solander, 1988). Nowadays, particular attention is paid to the relative role of aquatic vegetation, which influences physical, chemical and biological parameters.

Macrophytes or water vegetation can have opposite impacts on the water body. Once a lake has lost its submerged plants, it can be very difficult to re-establish them. This is thought to be due to 'alternative stable states', the idea that shallow lakes exist in either a state dominated by plants or a state dominated by phytoplankton (microscopic algae). Each state has a set of feedbacks that operate to buffer against changing to the other state. The switch from a plant-dominated lake to a phytoplankton-dominated lake most often occurs as the lake becomes enriched by nutrients, particularly nitrogen and phosphorus. The phytoplankton state is often associated with poor water quality, an unappealing 'pea-soup' appearance, and a loss of biodiversity as plant species become fewer, along with the animals that depend on them. Unfortunately, reducing the nutrient load of the lake doesn't necessarily result in a switch back to the more desirable plant-dominated state. This is because restoration of the necessary species of macrophytes in shallow lakes depends on several conditions;

Water clarity

Light is fundamental for plant growth. A decline in water clarity due to increased matter in the water column (such as particles, phytoplankton, dissolved organic matter) limits the depth to which plants can grow. However, water clarity can improve over time in some situations, such as when catchment initiatives are undertaken.

Plant inocula

There need to be inocula (seeds or plant fragments from other areas) that can respond to improving conditions in a lake. Seed banks that are old or depleted and have a poor germination response can be a barrier to plant restoration.

Wave action

Wind-driven wave re-suspension of bed sediments can be a barrier in shallow lakes with large fetches, where plants are uprooted, buried, or shaded out by high levels of re-suspended matter in the water column.

Pest fish

Certain alien fish can adversely affect plants by eating, damaging, or uprooting them and by promoting turbid conditions through foraging in lake bed sediments.

Catchment initiatives

Various measures can enhance a lake's chance of regaining vegetation. These include sediment traps or buffering vegetation along streams and lake edges to improve clarity and water quality.

In this research the focus is on shallow lakes of the type "Shoreline" brackish water in which there are different aquatic species dominating in both the littoral and limnitic zones. Despite major disturbances, the Delta lakes, which are of this type, are still primarily characterized by extensive stands of emergent macrophytes (Khedr & Lovett, 2000) that form valuable refuges for wild-life. At *Edko Lake*, the dominant emergent plants were *Phragmites communis* and *Typha domingensis*. *T. domingensis* was frequently present as fringing growth along the edges of the main *Phragmites* stands. Frequent channels enclosed by emergent vegetation were colonized by water hyacinth *Eichhornia crassipes* and *Potamogeton pectinatus*. Patches of *P. pectinatus* were occasionally common beyond the reed beds. This plant species was formerly much more common (Farinha *et al.*, 1996). Nowadays it still exists around the drain inlets into the lake and close to the south-north and south-east shores of the lake. A detailed description of the vegetation types and their photos is given in Chapter 6.

Fisheries (Aquaculture)

Various types of aquaculture form an important component within agricultural and farming systems development. These can contribute to the alleviation of food insecurity, malnutrition and poverty through the provision of food of high nutritional value, income and employment generation, decreased risk of monoculture production failure, improved access to water, enhanced aquatic resource management and increased farm sustainability (e.g. FAO 2000a, Prein and Ahmed 2000). Inland fisheries and fish ponds surrounding lakes and drainage systems within agricultural areas, may yet be able to yield more fish as effort increases, but the increased effort required will become increasingly challenging. Inland fisheries are also vulnerable to environmental impacts, such as watershed degradation, development of water control structures and pollution. All features of the changing rural environment. Thus, aquaculture has an important role to play in meeting the increasing demand for fish. Indeed, the growth of global aquaculture is forecasted to continue for some time (FAO 2000c). The potential negative impact due to miss-planned and uncoordinated development of these fisheries in developing countries is quite obvious. Depending on how it's done and what species are farmed, inland fisheries or fish ponds development has the potential to cause many problems. These include:

- Pollution within the fish ponds from wastes such as particulate matter from faecal material and uneaten food, nutrients, and chemicals and drugs, such as pesticides, disinfectants, and antibiotics.

- Transfer of pollution to the surrounding water bodies, specifically lakes and wetlands, due to continuous interaction and discharge of wastes from the ponds to the water bodies.

- Negative impacts on lakes and surrounding water bodies of eutrophication due to excess nutrients loadings.

- Negative impacts on wild populations of fish through the escape of farmed fish and the transfer of diseases and parasites to open water bodies, as well as negative impacts on other wildlife.

- Privatization of the fisheries and negative interactions with other stakeholders.

Eutrophication Levels and Indicators

Their usual location on the flatter, agricultural lowlands makes shallow lakes more vulnerable and they have become turbid during this century due to eutrophication. The response of shallow lakes to eutrophication is often catastrophic rather than smooth transition by the complete loss of submerged plants, which are the most essential component of a pristine state of such lake.

In its simplest expression, eutrophication is the biological response to excess nutrient inputs to a lake. The increase in biomass results in a number of effects which individually and collectively lead to impaired water use. Meybeck *et al.* (1989) highlight that eutrophication is a natural process which, in many surface waters, results in beneficial high biomass productivity with high fish yields. Accelerated, or human-induced, change in the trophic status above the natural lake state is the common cause of the problems associated with eutrophication. Such human induced changes may occur in any water body, including coastal marine waters, although the progression and effects of eutrophication are also mediated by climate. As a result warm tropical and sub-tropical lakes are more severely affected than colder lakes.

High nutrient concentrations in a lake are derived from external inputs from the watershed. The final biomass attained is determined primarily by the pool of nutrients available for growth at the beginning of the growing season. The primary nutrients, such as nitrogen and phosphorus, are used until growth is complete and the exhaustion of the pool of either one of them places a final limit on the phytoplankton growth. By definition, the nutrient which is exhausted is the limiting nutrient in any lake system. Meybeck *et al.* (1989) suggest that, in waters with an N/P ratio greater than 7 to 10, phosphorus will be limiting, whereas nitrogen will be limiting in lakes with an N/P ratio below 7.

Changes in transparency in lakes may be caused by increasing turbidity due to increasing concentrations of mineral material or to increasing plankton biomass. Increases in mineral matter are caused by:
- Turbid in-flow in fluvial waters with high watershed erosion rates,
- Re-suspension of bottom sediment by wave action in shallow lakes, or shallow areas of lakes (when wave height and wavelength developed during storm events allow direct interaction with the bottom to take place), and

- Shore line erosion due to wave impingement and to surface water gully erosion of unconsolidated shore line material.

Lakes and other water bodies depend mainly on their watersheds for nutrients and other substances to sustain biological activities. While these nutrients and substances are required for a healthy aquatic environment, an excess of these inputs leads to nutrient enrichment and eutrophication of the lake. Eutrophication of a water body is usually quantified in terms of the concentration of the CHL-aorophyll contained in the algal/plankton. While aging of water bodies is a normal process, detailed information on eutrophication (CHL-aorophyll content) in a lake will allow better management plans to be developed to control the source in nutrient input from the watershed.

The sequence of changes in shallow lakes during eutrophication is not well documented, but some elements are agreed upon by most workers in the field (Moss, 1988). Shallow lakes with a low nutrient content usually have vegetation dominated by relatively small plants. With increased nutrient loading the biomass of aquatic macrophytes increases and plants that fill the entire water column or concentrate much of their biomass in the upper water layer become dominant. Such dense weed beds are often experienced as a nuisance by the fishing and boating public. When weed control programs eradicate the vegetation, turbidity in shallow lakes tends to increase strongly due to algal blooms and wind re-suspension of the sediment. Also, when vegetation is not controlled explicitly, further eutrophication of vegetated lakes can lead to a gradual increase of phytoplankton biomass and of the periphyton layer that covers the plants. Shading by these organisms ultimately leads to a collapse of the vegetation due to light limitation.

CHL-aorophyll-a concentration is considered an indicator for the trophic condition of a lake. CHL-aorophyll-a is a green pigment found in plants. It absorbs sunlight and converts it to sugar during photosynthesis. CHL-aorophyll a concentrations are an indicator of phytoplankton abundance and biomass in coastal and estuarine waters. They can be an effective measure of trophic status, they are potential indicators of maximum photosynthetic rate and are a commonly used measure of water quality. High levels often indicate poor water quality and low levels often suggest good conditions. However, elevated CHL-aorophyll-a concentrations are not necessarily a bad thing. It is the long-term persistence of elevated levels that is a problem. *For this reason, annual median CHL-aorophyll-a concentrations in a waterway are an important indicator.*

Changes to systems which decrease (e.g. construction of canal estates) or increase (e.g. breakwaters, training water and dredging) flushing rates influence CHL-aorophyll-a concentrations because flushing dilutes nutrients and moves them away from plants, making them less available. Conversely, slow moving or stagnant waters let nutrients increase and cell numbers grow. Elevated concentrations of CHL-aorophyll-a can reflect an increase in nutrient loads and increasing trends can indicate eutrophication of aquatic ecosystems.

Measures of eutrophication levels in lakes are used and based on concentration of nutrients. Numerous indices are available for evaluating the trophic state or nutrient condition of lakes. Among the more widely used is Carlson's Trophic State Index (TSI) (Carlson 1977). Carlson TSI values can be derived for a water body from any of 3 water quality parameters: water clarity (*Secchi disk depth*), *CHL-aorophyll-a concentration*, and *total phosphorus concentration*. TSI values represent a continuum ranging from very clear, nutrient-poor water (low TSIs) to extremely productive, nutrient-rich water (high TSIs). The Carlson TSI was based on a logarithmic base of 2; each doubling of algal biomass results in halving the Secchi disk reading. Because the scale was multiplied by 10, each increase of 10 TSI units corresponds to a doubling in algal biomass. Carlson's TSI scale and equations have been widely accepted and applied throughout North America in evaluating and comparing trophic condition of lakes. But these indicators should be tested in different geographic regions, as the parameters used in developing these indicators formulas can be site specific, based on applications done in the United States.

2.6.   WATER QUALITY MANAGEMENT: *TRENDS AND ISSUES*

As earlier above, the movement of freshwater is facilitated by the hydrological cycle (including floods and droughts) which demonstrates that water is inextricably linked with the environment. This implies that actions in the environment, and more specifically the catchment/watershed, impact on water resources. Similarly, water impacts on the environment. However the very nature of water is such that its characteristics vary in both quantity and quality in time and space. The quality of water itself is a significant factor to the limits on the amount of available water that may be used for various purposes.

What nature provides as available quantities may be reduced if the quality of water does not permit its use for a specific purpose. This leads to additional costs in the development of the water resource to meet both quantity and quality requirements needed for the particular use. Thus water quality directly affects the quantity of water that can be employed for various purposes. The rapid growth of population and urban communities and industrial and agricultural activities has significantly contributed to the deterioration of water quality of lakes, rivers, groundwater and coastal waters worldwide. Pollution of the environment including its freshwater resources is the result of human society's activities. Also the use of water results in its degradation. Often this is incremental and contamination results through repeated use.

It is expected that with economic development in the coming years, there will be a substantial increase of water demand and therefore an increase of pollution discharge, if measures are taken. Already as populations and economic activities grow many countries are rapidly reaching conditions of water scarcity or facing limits to economic development. The widespread scarcity, gradual degradation, and pollution of freshwater resources in many regions of the world, along with the encroachment of incompatible activities, demand integrated water resources planning and management. It is therefore of great importance that freshwater as a finite and vulnerable resource should be managed holistically taking into consideration the integration of sectoral water plans and programmes within the framework of national economic and social policy. Management must also recognize the interconnection between water bodies and of the elements related to freshwater and freshwater quality.

Water quality management deals with all aspects of water quality problems relating to the many beneficial uses of water. Water quality is a reflection or response of water composition to all inputs and processes, whether natural or cultural. Water quality management should not be equated to water pollution control which generally is the adequate treatment and disposal of wastewater. In the definition of water quality management, water uses consist of intake, on-site, and in-stream flow uses. Intake uses include water for domestic, agricultural and industrial purposes, or uses that remove water from the source. On-site uses primarily refer to water consumed by swamps, wetlands, evaporation from water bodies, natural vegetation, and unirrigated crops and wildlife. Flow uses include water for estuaries, wastewater dilution, navigation, hydroelectric power production, and fish, wildlife and recreation purposes. Water quality management serves to optimize water quality for all beneficial uses. Implied in this, is that water should be managed so that use at any one location will not be detrimental to its use at another location. In managing water quality, the factors and inputs that must be considered, include both man-made sources and natural causes. With respect to water quality changes, natural causes include geologic formations, vegetation, geographic factors and natural eutrophication. Water quality is assessed by its physical, biological and chemical characteristics. Contamination can alter one or all of these characteristics and may originate from point or from ambient sources. The investigation and management of water resources systems for water quality must include consideration and evaluation of;

(a) The physical, chemical and biological composition of headwaters and significant groundwater discharges.
(b) Water quantity and quality requirements for all existing and potential water uses
(c) The means of water withdrawal and their effect on water quality and quantity
(d) The existing and future water and wastewater treatment technology used to alter water quality

(e) The wastewater outfall configuration and effluent mixing
(f) The eutrophication status of the receiving waters
(g) The waste assimilative capacity of the receiving waters.
(h) The ecological changes that might be caused by wastewater discharges
(i) The potential effects of discharged waters.

Major problems affecting the water quality of rivers and lakes vary according to the specific situations as mentioned earlier. Problems may arise from inadequately treated sewage, poor land use practices, loss and destruction of catchment areas, inadequate controls on the discharges of industrial waste waters, poor siting of industrial plants, deforestation, uncontrolled poor agricultural practices and a lack of integrated watershed management. Some of the effects of the above are leaching of nutrients and pesticides, threatened ecosystems, public health risks, erosion, sedimentation and deforestation leading to land degradation. Many of these negative effects may have arisen from environmentally destructive development and a lack of public awareness and education on the protection of surface and groundwater resources.

To effectively control water quality, it must be described in **precise technical quantitative terms** to allow the decision for effluent discharge limitations or a beneficial use of the water to be formulated. In addition any requirements for water quality must be imposed with consideration of the concomitant level of treatment requirements of wastewater effluents or water supply intakes upstream and downstream from the point of interest.

Water Quality Management in Developing Countries

The water quality situation in developing countries is highly variable reflecting social, economic and physical factors as well as state of development. And while not all countries are facing a crisis of water shortage, all have to a greater or lesser extent serious problems associated with degraded water quality. In some countries these are mainly associated with rivers, in others it is groundwater, and in yet others it is large lakes; in many countries it is all three. Because the range of polluting activities is highly variable from one country to another, and the nature of environmental and socio-economic impacts is equally variable, there is no "one-size-fits-all" solution. There are, however, some common denominators in the types of actions that are required for sustainable solutions (Ongley, 2000).

For the purpose of quantifying the water quality, numerical values of the concentration of various substances present in the water samples are determined through physical, chemical and biological techniques of analysis. Until today, in most of developing countries, the major quantification tend of the water quality, is based on a statistical evaluation of a number of samples taken at various locations, flows and times. Accuracy of the water quality quantification is related to the frequency of the data acquisition and its statistical reliability. The process of quantifying water then involves a comparison of the statistical water quality characteristics with water quality criteria or standards. Water quality standards used throughout the world are generally either in the category of stream standards or effluent standards or a combination. In the United States, although primary emphasis is on effluent standards, stream standards are enforced where the effluent loadings exceed the waste assimilative capacity, as determined by stream standards (water quality limiting cases).
Apart from effluent regulations and, sometimes, national water quality guidelines, a common observation is that few developing countries include water quality within a meaningful national water policy context. Whereas water supply is seen as a national issue, pollution is mainly felt at, and dealt with, at the local level. National governments, with few exceptions, have little information on the relative importance of various types of pollution (agriculture, municipal, industrial, animal husbandry, aquaculture, etc.) and therefore have no notion of which is of greatest economic or public health significance. Usually, freshwater quality management is completely divorced from coastal management even though these are intimately linked. Consequently, it is difficult to develop a strategic water quality management plan or to efficiently focus domestic and donor funds on priority issues.

Data is always the key issue for the design and implementation of management information systems. The first design criterion in any water quality program is to determine *the management issues for which water quality data are required.* The technical aspects of data collection will flow from this decision, especially as there are now very cost-effective alternatives to conventional monitoring practice. Establishing of data objectives in Mexico, for example, resulted in a radical shift in national monitoring practice which produced the savings in the funds dedicated to monitoring networks. Also, these new methods will permit a much higher level of regulatory compliance. Most importantly, data programs are now seen to have value insofar as they will provide a service for someone other than the monitoring agency itself.

Most developing countries are "data-poor" environments as well as being challenged by economic restrictions. This, together with lack of sufficient technical and institutional capacity and often a poor scientific knowledge base, suggests that the conventional "western" approach to water quality monitoring and management is not well suited to many if not most developing countries. *It is, therefore, timely to promote a new water quality paradigm that is more suitable, affordable, and sustainable in developing countries.* For more advanced developing countries or where there are issues such as contamination from point and non-point sources, the conventional and expensive chemical approach to monitoring can be effectively replaced by new diagnostic tools such as diagnostic chemistry and biological assessment. While these never completely replace bench chemistry, the trend is to use these inexpensive diagnostic tools to determine whether or not the pollutant load meet certain predetermined levels of risk before any expensive chemistry is performed.

Another area of technical innovation that has considerable merit in developing countries is the application of new decision-support (DSS) capabilities drawn from the field of information technology (IT). These techniques are particularly useful in data-poor environments that are typical of developing countries. There is a large knowledge-base (*domain knowledge*) in the scientific community on most types of water quality management issues which, when supplemented by *local knowledge*, can greatly facilitate decisions on water quality management. The objective of a well–designed decision support system (DSS) is to put domain knowledge into the hands of local practitioners in such a way that the user is guided through a complex task to a conclusion for which the results can be expressed in degrees of confidence. Although decision-support technology is now well known, there has been little effort by the international community to systematically develop these technologies and related data and knowledge bases so that these can be applied to typical water management issues in data-poor or knowledge-poor environments.

Existing Management Issues and Difficulties

Since the intended use of water dictates the water quality requirements, delineation of water allocated for specific uses is mandatory. The traditionally accepted beneficial uses of water reflect the multi-interest utilization of water resources. These uses include domestic water supply, industrial, water supply, agricultural water supply, fisheries, urban development, hydropower generation, transportation (navigation), recreational waters, sanitation, assimilation of wastes and other activities. Clearly these uses span a wide spectrum of water quality requirements. In developing countries, a key management problem in this context is the lack of dedicated monitoring objective for each of these different uses of water.

Monitoring networks and data gathering and dissemination systems are not designed based on clear information needs and priorities but rather on available tools and capacities. This has led to 'monitoring for monitoring's sake' and 'data rich but information poor' situations. Most developing countries have initiated data collection and water quality monitoring. Many different institutes are involved in water quality monitoring and carry out their assigned mandate for their own objectives, independent of other related activities. For example, in Egypt, the Ministry of Health monitors water intake, the Ministry of Water Resources and Irrigation monitors ambient water quality, and the Ministry of Environment and the Ministry of Health both monitor water quality for compliance testing (Abdel-Dayem, 1994). There is no cross-ministerial mechanism to review policies and mandates, provide for monitoring coordination, nor facilitate the exchange of information and data

The identification and categorization of water quality problems does not always take place in appropriate and planned measures. Identified water quality problems may fall into different categories requiring application of different management tools and interventions for optimal resolution of the problems. For example, it is important to know whether a certain water quality problem pertains only to a local community or whether it is a national problem. If a problem exists at the national scale it might be necessary to consider imposing general effluent standards, regulations or other relevant measures, also the type of management tools to be applied will vary. By contrast, if the problem is limited to a small geographic region it might only be necessary to consider issuing a local by-law or to intervene to settle a dispute through mediation.

It may also be useful to categorize water quality problems as either "impact issues" or "user-requirement issues". Impact issues are those derived from human activities that negatively affect water quality or that result in environmental degradation. User requirement issues are those which derive from an inadequate matching of user specified water quality requirements (demand) and the actual quality of the available resources (supply). Both types of issues require intervention from a structure or institution with powers that can resolve the issue in as rational a manner as possible, taking into consideration the prevailing circumstances. According to the traditional water pollution control approach, user-requirement issues would often be overlooked because the identification of such problems is not based on objectively verifiable indicators. Whereas an impact issue can be identified by the presence of, for example, a pollution source or a human activity causing deterioration of the aquatic resources (e.g. eutrophication), user-requirement issues are identified by a lack of water of adequate quality for a specific, intended use.

In many countries, no comprehensive and coherent policy and legislation exists for water pollution control or for environmental protection. This does not prevent water pollution control from taking place before such policies have been formulated and adopted, but the most efficient and effective outcome of water pollution control is obtained within a framework of defined policies, plans and co-ordinating activities. There may be obvious shortcomings in the existing situation that need urgent attention and for which remedial actions may be required independently of the overall general policy and planning. Such interventions and remedial actions should be taken whether or not an overall policy exists. A lack of policy should not delay the implementation of identified possibilities for obvious improvements in water pollution control. The only means to develop these policies and take decisions for remedial action, is having a reliable and adequate decision support tool (DSS) that is based on using appropriate management tools. Identification of the existing water quality problems and understanding the behavior of water systems under the impact of these problems is an important need before policy and decision s can take place. Nowadays, with the exceeding and complex water quality problems, and with vast developments in data acquisition and modeling tools, it is not sufficient to deal with water quality quantitative information in the form of statistical analysis or trends.

Modelling tools are treated here as any set of instructions based on a deterministic theory of cause-effect relationships which are able to quantify a specific water quality problem and thereby support rational management decisions. This can be done at different levels of complexity, some of which are discussed below form the application point of view:

- *Loadings:* Preliminary decisions can be taken with respect to reduction of loadings from a ranking of the size of actual pollution loadings to a particular receiving water body. The rationale is to assess where the greatest reduction in pollution can be obtained in relation to the costs involved.

- **Mass balances:** Mass balances can be established using load estimates from pollution sources in combination with the water flow or residence time in the water body. The significance of the different loadings can be evaluated by comparing their magnitude to their contribution to the resulting concentration of the pollutant in the receiving waters. The significance of the different loadings for the pollution level of the receiving water body provides the rational basis for decisions on effective reduction of the pollution level in those waters.

- **Effect evaluation:** Assessment of changes in the identified pollution sources and their resulting concentration in the receiving waters can be made at various levels, from using simple, empirical relations to long-term mass balance models. An example of a well known empirical relation is the Vollenweider method for estimating eutrophication effects in lakes (Vollenweider, 1968, 1975, 1976). Based on experience from measurements in a large number of lakes, the method relates pollution discharges and static lake characteristics (such as water depth and retention time) to expected effects on the Secchi depth and algal concentrations. Effect evaluation may also combine considerations about cost effective pollution reduction at the source, the resulting pollution concentration in receiving waters and the resulting effects in the ecosystem.

- **Simple mathematical mass balance models:** Application of this tool allows consideration of the possible changes over time in relation to any reductions proposed in pollution load. Many types of these biogeochemical models have been developed over the years and some are available in the public domain.

- **Advanced ecological model:.** If higher level effects of pollution loadings on an ecosystem are to be determined, more sophisticated ecological models are available. Such models may create the basis for a refined level of prediction and should be used in cases of receiving waters with high complexity and importance, provided sufficient resources (financial, human or institutional) exist or can be allocated.

The above examples serve to illustrate that quantitative assessment of pollution problems can be performed at various levels of complexity, from hand calculations and statistical analysis to advanced state-of the-art ecological modelling.

There is a wide range of existing water quality management tools. The range of tools and instruments should be considered as an input to the overall process of achieving effective water pollution control. They are considered necessary means to address the identified problems. The manager's task is to decide which tool(s) will most adequately solve the present water pollution problem and to ensure that the selected tool(s) are made available and operational within the appropriate institutions. Integration of tools is considered important in some situations where there is a need for better understanding of the problem and clear interpretation of existing situation.

(a) Regulations, management procedures and by-laws
(b) Water quality standards
(c) Economic instruments
(d) Monitoring systems
(e) Water quality modelling tools
(f) Environmental impact assessment and cross-sectoral co-ordination

The key actions to address water quality management issues and problems include;

1. Development of appropriate, cost effective and reliable data programmes that can inform sound judgments on environmental policy, management and regulatory needs. This is most important when we realize that developing countries are extremely data poor, both in amount and reliability and that part of the current water crisis is the failure of national data programmes to produce information that can be used to estimate the contributions and impacts of different categories of

pollution sources. Development of the appropriate management systems depends very much on the quality and availability of the right data.

2. Water resources protection and conservation. One focus of this must be the rehabilitation or remediation of important but degraded catchment areas as a means of increasing the quantity of useable water both for human and ecosystem needs.

3. Water pollution prevention and control. In this there is the application of the 'polluter pays' principle to the sources where appropriate, establishment of standards for the discharge of effluents and for the receiving waters, the use of new technologies, product and process change, pollution reduction at source and effluent reuse, recycling and recovery, treatment, and environmentally safe disposal for pollution minimization. In addition, there is the mandatory environmental impact assessment of major water resource development projects with the potential to impair water quality and ecosystems. Also included are the identification and application of best environmental practices at reasonable cost.

4. Development and application of clean technology. This focuses on the control of industrial waste discharges in an integrated manner and through the application of measures derived from a broad based life cycle analysis. Also included are the treatment of municipal wastewater for safe reuse in agriculture and aquaculture, the development of biotechnology for waste treatment, and the development of appropriate methods for water pollution control considering traditional and indigenous practices.

5. Protection of aquatic ecosystems. This focuses on the rehabilitation of polluted and degraded waters to restore aquatic habitats and ecosystems and to conserve and protect wetlands.

6. Protection of freshwater living resources. This focuses on the control and monitoring of water quality to allow for sustainable development of inland fisheries and the protection of ecosystems from pollution and degradation. Monitoring and surveillance of water resources and waters receiving wastes. This focuses on the establishment of networks for monitoring of waters receiving wastes and pollution, surveillance of pollution sources to improve compliance, the application of environmental impact assessment and geographic information systems, and the monitoring of chemical utilization and national land use to prevent degradation.

7. Development of legal instruments to protect the quality of water resources. This focuses on the monitoring and control of pollution and the application of environmental impact assessment.

3. INTEGRATING TOOLS FOR SURFACE WATER QUALITY MANAGEMENT

This chapter focuses on the main used tools in this research study. It gives a general overview of the different tools used in building the water quality management information system. It highlights the importance of monitoring and data acquisition techniques then it explains the different types of hydrodynamic and water quality models. The importance of remote sensing as a tool for water quality analysis and management is also shown. The chapter highlights the role of information systems in water quality management and the use of decision support systems in the management process. It highlights the different uncertainties associated with the decision support tools and their uses in water quality management.

3.1. MONITORING AND DATA ACQUISITION TECHNIQUES

The ultimate goal of monitoring is to provide information, not data. In the past, many monitoring programmes have been characterised by the "data rich, information poor syndrome" (DRIP syndrome; Ward et al., 1986). There should be more attention on the analysis and further use of collected data so that the end product of monitoring is information. Another notion that became obvious over the last decade is that the connection between the data collected by monitoring and its use within management and policy is an important element in the success of any (water quality) monitoring system design. Water quality monitoring (information) systems should be a balanced combination of data collection and information generation. This is illustrated in the monitoring cycle, Figure 3-1, which illustrates that monitoring is a sequence of related activities that begins with the definition of information needs and ends (and starts again) with the use of information products. Too often water monitoring has been viewed as only the first three steps listed above. In other words, once data are stored in a computer, the monitoring task is completed. Data are hereby viewed as the final "product". In order to make monitoring systems more effective and efficient there is a large concern making monitoring tailor-made (e.g. Adriaanse et al., 1994; Ottens et al., 1997) or optimising monitoring. Tailor-made implies 'providing the right information to make management decisions' (Adriaanse, 1995).

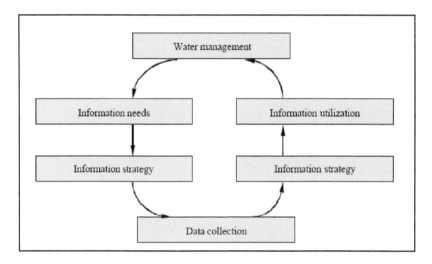

Figure (3-1): The information cycle (after Timmerman, 2000)

According to (Massdam, 2000), water monitoring systems have evolved through several phases:

1. No formal water quality data collection or information generation

2. *Monitoring Networks:* Early twentieth-century water quality data collection networks (U.S.A) were designed to collect data to answer narrow technical questions with the respect to limited areas and limited time periods. Many early efforts focus on the prevention of nuisances. Over time, monitoring network design embodied improved data collection and storage techniques, and an increased appreciation of information needs and the stochastic nature of water quality. The focus began to shift to pollution control, especially when the relationship between water quality and public health began to influence legislation. Monitoring networks really taking a flight when laws on the water pollution are placed in action (USA 1940-1950, The Netherlands 1970). Much of the monitoring efforts is to determine whether water quality is sufficient for a set of designated uses. Samples were collected, checked against criteria for an established use, and determined to

be acceptable or non-acceptable. Schemes show a more top-down sequence of a restricted number of activities, starting from a rather arbitrarily chosen network and having an open end concerning the production of data.

3. *Monitoring Systems:* The 1980s witnessed continuing development of data collection and storage methods, refined statistical characterisation of water quality variable behaviour (especially directed toward trend-detection) and a merging appreciation of the need to design water quality networks according to systems principles. The distinction between data and information was emphasised. Systematic design procedures developed included a 12-Step Process (Sanders et al., 1979), a 5 Step-Process (Sanders et al., 1987), a formal (4-task) Systematic Design Procedure (Mar et al., 1986), the New-Zealand National Network design (Smith et al., 1989), a 5 step (optimisation) process (Schilperoort and Groot, 1983) and "wheel and axle" framework (Payne and Ford, 1988).

4. *Information Systems:* The late 1980's and early 1990s witnessed continued development of data collection and statistical analyses, and of frameworks to guide the design of more comprehensive water information systems. The guiding principle in this is that monitoring (and assessment) should be seen as a sequence of related activities that starts with the definition of information needs and ends with the use of the information products. This cycle of activities is called the "monitoring cycle" (Figure 6). Increasingly, their potential contribution to broad water management systems and ecological management systems has been recognised. To begin to address the problem, the concept of a "*water quality information system*" was proposed by Ward (1979) and Ward (1986). Examples of these efforts include a "5 step framework for designing water quality information systems (Ward et al., 1990) and the development of "Data Analysis Protocols" (Atkins, 1993).

5. *Management Information Systems:* Recent water management trends have set the standard for a new era. Water monitoring and information systems must become Water Management Information Systems. As management information systems, these programs must:

    *   Derive from goals and associated water management objectives
    *   Satisfy expanded water management objectives and information needs.
    *   link directly to specific water management decision needs and decision processes ("monitoring for action")
    *   Be dynamic, i.e. designed to be continuously reviewed and improved (Hotto et al., 1997; Misseyer, 1999).

The scope of this study is beyond detailed discussion on the water quality monitoring but it is an important part to be highlighted. To deal with water quality modeling and integrating it with other tools for management purpose, it is crucial to point to the importance of data acquisition and availability with a certain quality, so eventually it could be transferred from monitoring to action by means of management tools. Evaluating water quality in a meaningful way is a complex issue and involves several items to be taken into consideration. A single sample from a specific water body can only tell what the water quality is at that particular time and that specific location. This is true because of changes that occur in water chemistry as water moves downstream, as evaporation from the soil surface occurs, and as transpiration (the uptake and loss of water) by plants takes place. In order to develop an accurate understanding of how clean or polluted a particular water body is, especially the flowing streams, rivers and lakes, it is necessary to collect and analyze water samples from the same location over time and in different locations over different times. Another important factor in evaluating water quality data is the need for measurement of the flow rate and the dimensions and physical properties of the water body. Climate and meteorological data are also of great importance for detailed assessment studies of water quality. Therefore, water quality assessment methodologies involve data acquisition in different categories and forms.

Traditionally, monitoring programmes collect data either from chemical and biological analysis of water samples or from field equipment. However, depending on available laboratory facilities, instruments transport and human resources, for example, all monitoring programmes are restricted in some way and may collect data primarily by direct sampling. A number of information gaps often have to be filled, therefore, before a rational decision about monitoring system design can be taken with respect to a specific water quality problem. Although they are less accurate, indirect techniques for obtaining the necessary information exist for a variety of water quality-related factors. It is possible, for example, to obtain reasonable estimates of pollution quantities from various sources from knowledge of the activities causing the pollution. In some catchments where the monitoring does not cove large areas, indirect estimation of pollutants of loads based on representative measurements of similar catchments could be possible.

Another frequent problem associated with traditional monitoring programmes is the lack of coupling between measured concentrations and water flow or discharge measurements, thereby rendering quantification of pollution transport difficult. Estimation techniques also exist for these situations, where hydrometric networks are not established or functioning, or where instruments are not available for measuring flow, such as in wastewater discharges. The actual design of a fully operational and adequate national monitoring system must, from the beginning, take account of the requirements of the additional management tools which are being considered for use. The complexity and size of the area to be monitored, the number of pollutants monitored, and the frequency of monitoring, have to be balanced against the resources available for monitoring. To a large extent the data that become available determine the level of complexity of the management tools that can be supported by the monitoring system.

Monitoring programmes should have clear objectives and ultimate goals for data acquisition. There should be a long term vision for monitoring in addition to traditional assessment of water quality problems and calculation of statistical trends in water quality. Since there are some limitations in monitoring programmes related to time and space variations, if the objectives of monitoring are linked to the management tools that are used such as mathematical models, the programs could be adjusted to serve the use of these tools. Also this would clarify where could be the gaps in data acquisition techniques and could introduce the use of complementary tools to enhance the management process for example the use of remote sensing as a data acquisition tool.

The integration in monitoring programmes is needed to establish management plans on a watershed scale, for example the monitoring of catchment water quality should not be separated from lakes monitoring programmes. A common observation of lakes water quality programmes is that they tend to be inefficient, the data are of uncertain reality if it exist at all in some locations, programme objectives are poorly linked to management needs for data, the analytical technology is often old and inefficient, focus is on water chemistry even though water is known to be a poor monitoring medium for many toxic chemicals, and data bases are incapable of mobilizing for management purposes (Ongley, 1993).

3.2. GIS APPLICATIONS IN WATER QUALITY: DATA HANDLING, PROCESSING AND MODELLING

To avoid the "data rich but information poor" syndrome, data analysis, information generation and reporting should be given the same attention as the generation of the data themselves. Water pollution control requires access to statistical, graphical and modelling tools for analysis and interpretation of data. Data used for water pollution control, such as water quality, hydrology, climate, pollution load, land use and fertiliser application, are often measured in different units and at different temporal and spatial scales. In addition, the data sources are often very diverse (Demayo and Steel, 1996). Most published definitions of "geographic information systems" refer to both data and operations, as in "a system for input, storage, manipulation, analysis and output of geographical referenced information". In turn, geographical referenced information can defined fairly robustly as information linked to specific locations on the Earth's surface. To obtain information about, for example, spatial extent and causes of water quality problems (such as the effects of land-use practices), computer-based GISs are valuable tools. They can be used for data presentation, analysis and interpretation. Geographical information systems allow the georeferencing of data, analysis and display of multiple layers of geographically referenced information and have proven their value in many aspects of water pollution control.

GIS uses a computer database to store large quantities of spatial and temporal data. This allows the integration of diverse types of information into a form that makes it possible to consider different approaches to land management and environmental problems before making management decisions. Spatial data is information that describes how a specific feature is located or distributed in space. This type of information can include watershed boundaries, slope, aspect, contour, soil type, stream location, waterbodies boundaries and land use/land cover. The use of GIS allows us to process and evaluate these data. Without this type of computer tool, such large amounts of data would be very difficult to interpret. Information stored in a GIS will come from a variety of sources. The greater the quantity and quality of the information, the more complete the GIS database will be.

"GIS" at its broadest has become a term to refer to any and all computer-based activities that focus on geographic information; "GIS data" is often used as shorthand for digital geographic information; and the redundant "GIS system" is becoming the preferred term for the software itself (Goodchild, 1999). The "GIS system" requirements of most (current) monitoring systems are, in fact quite small (usually limited to handling georeferenced site information, spatial mapping, and limited map overlaying). If we go into broader types of application, GIS is used as a link between data sets and modeling tools or it is even used as a standalone modeling tool in some water quality assessment studies. The general believe in water management is that GIS systems are only for specialised activities because the learning requirements are substantial, the hardware and software cost are high and only specialists can efficiently use such systems (Ongley, 1997). But nowadays several studies and conducted research shows the extensive use and the importance of this tool in specialized analysis and detailed studies in the field of water in general and in the environmental and water quality studies in specific.

For example, GIS as a standalone tool have been used to provide water quality information on:

- Location, spatial distribution and area affected by point-source and non-point source pollution.
- Correlations between land cover and topographic data with environmental variables, such as surface run-off, drainage and drainage basin size.
- Mapping of water quality parameters.
- Presentation of monitoring and modelling results at a geographic scale.

A typical GIS system consists of:
- A data input system which collects and processes spatial data from, for example, digitised map information, coded aerial photographs or satellite images, and geographically referenced data, such as water quality data.
- A data storage and retrieval system.
- A data manipulation and analysis system which transforms the data into a common form allowing for spatial analysis.
- A data reporting system which displays the data in graphs or maps showing spatial reference of information.

The application of GIS as a processing tool for remote sensing data will be explained also in the coming sections of this chapter.

3.3. HYDRODYNAMIC-WATER QUALITY MODELLING APPROACHES FOR CATCHMENT – SHALLOW LAKE SYSTEMS

Models are invaluable tools for resource management. Models help resource managers develop a shared conceptual understanding of complex natural systems, allow testing of management scenarios, predict outcomes of high risk and high cost environmental manipulations, and set priorities. Using mathematical models is a key for better understanding of natural systems specially water systems and complex water resources problems. Nowadays, relative to the present advances in computational sciences, hardware and software, improvement in rivers, catchments and lakes modeling has been only modest since the last decades. One of the reasons can be attributed to that in most of catchments modeling development has been undertaken by environmental scientist and engineers with limited computer science training. Most of the models they develop rarely make use of the power and efficiency of modern software engineering methods. Due to this, most catchments models are difficult to understand, use and adapt, leading to frustration amongst the users of such models and their developers (Argent et al., 2001). Environmental modelling is a specialist field, and different modelling approaches are specialist areas in themselves. Different philosophies abound; there are experts who advocate systems approaches using conceptual models and others who dismiss these as being too uncertain and based on opinion rather than fact. Even when the approach is agreed, experts may be at odds over which modelling product is superior.

Sometimes the needs of the resource manager appear to be lost in the technical debate between modellers who support fundamentally different approaches. The tensions between resource management and science are explored by Cullen (1990) and his findings are especially applicable to environmental modeling which can be expensive and highly technical. The quality of a water body as a part of the ecosystem, is reflected in its physical condition which strongly influences the chemical and biological processes that occur in water. Water quality interactions are by necessity simplified descriptions of an aquatic ecosystem that is extremely complex.

An essential tool, which can be used by water quality researchers and managers in developing management plans for rivers and watersheds, is mathematical or water quality modelling. Water quality modelling has had a long history (Streeter and Phelps, 1925, Velz, 1938, O'Connor, 1960, 962, 1967, Chapra, 1996). The increased computational power of modern computers, as well as the need for watershed or basin-wide planning, has resulted in the expansion of mathematical modelling codes beyond simple steady-state, one dimensional models of pollutants to complex, time-variable, three-dimensional models of nutrients / eutrophication and toxic materials. These models are often coupled to hydrodynamic and watershed or urban runoff models. Mathematical models can be used to predict changes in ambient water quality due to changes in discharges of wastewater. Predicting the water quality impacts of a single discharge can often be done quickly and sufficiently accurately with a simple model. Watershed water quality planning usually requires a model with a broader geographic scale, more data, and a more complex model structure. A water quality model is a set of mathematical equations that represent the physical, chemical, and biological characteristics and processes of a water body in a way that approximates reality. Water quality models can be used to simulate water quality changes that could be expected to result from changes in pollutant loads and different water quality management strategies. These simulations, called "scenarios," allow us to predict positive or negative changes on water quality.

3.3.1. Hydrodynamic - Water Quality Modelling Tools

Flow of water in a river or stream is described by the continuity and momentum equations. The latter is known as the Navier- Stokes or Reynolds equation. The actual form of a hydrodynamic model depends on assumptions made on characterizing turbulence. Methods vary from the use of eddy viscosity as known parameters to the application of the so called k-ε theory, (Bedford et al., 1988; Rodi, 1993). Complex models are available, (Abbott, 1979; Naot and Rodi, 1982) but for water quality purposes mostly the well-known, cross-sectionally integrated (1D) Saint Venant equations or approximations to these equations are used (Mahmood and Yevjevich, 1975; Abbott, 1979).

Many different forms and approximations to the Saint Venant equations are known, depending upon whether the flow is steady or unsteady and which simplifications are made. Thus, for water quality studies often the equation of steady, gradually variable flow is employed (which may be further simplified to the so-called Manning equation. Unsteady models, which are based on the continuity and momentum equations, include the kinematic, diffusive, and dynamic wave approaches. The difference stems from simplifications of the latter equation: dynamic wave models solve the full equation, diffusive models exclude the acceleration terms, while kinematic models disregard also the pressure gradient term that is essential for the description of backwater effects. (Routing methods used widely in hydrology usually correspond to the last approach, (Mahmood and Yevjevich, 1975). The hydrodynamic equations are generally solved by efficient finite difference methods (Mahmood and Yevjevich, 1975). For water quality issues the acceleration terms in the momentum equation rarely play a significant role and the typical time scales are amplified by conversion processes.

In general, three types of models are significant in the investigation of the environmental impact of water bodies:

1) **Hydrodynamic models** which describe the velocity and salinity distributions within the study area.
2) **Water quality models** which predict physical characteristics and chemical constituent concentrations of the waiter at various locations within the study area.
3) **Ecological models** which predict the interactions between water quality and the aquatic community.

3.3.2. Data Requirements for Mathematical Modelling of Water Quality

As it might be expected, the data requirements for different models increase with the level of complexity and scope of application. All models require data on flows and water temperatures. Static, deterministic models require point estimates of these data and often use worst case "design flow" estimates to capture the behavior of pollutants under the worst plausible circumstances. For most management purposes, the worst case will be high summer temperatures, which intensify problems with dissolved oxygen and algal growth, and low flows, which lead to high concentrations of BOD and other pollutants. Dynamic models will need time-series data on flows, temperatures, and other parameters. In addition to hydraulic data, models require base case concentrations of the water quality parameters of interest (dissolved oxygen, mercury, and so on). These are required both to calibrate the models to existing conditions and to provide a base against which to assess the effects of management alternatives. The models also need discharges or loads of the pollutants under consideration from the sources (e.g., industrial plants) being studied. The types and amounts of data needed for a given application are specific to the management or research question at hand.

The information derived from hydrodynamic models forms the basic part of the database for water quality and ecological models, this data in turn will become necessary for water quality models that would then be part of the database for ecological models. Hence, it is essential that these foundation modeling activities be accomplished with adequate accuracy. The various models described require input data which may be classified as:

- Data that describe the initial conditions of the system.
- Data that describe the "boundary conditions" of the system These data include system geometry and the quantity and constituent concentration of freshwater inflows.
- Other data necessary for the calibration of the models, including a description of the hydrography of the study area.

As no model study can be more accurate than the information on which it is based, the importance of adequate field data cannot be overemphasized.

The first steps in any model study is to specify the objectives; an assessment of the geophysical, chemical, and biological factors involved; and collection of data essential to describe these factors. Data collection and assessment should include the following remarks:

- An identification of the various freshwater inflow sources, including their average, range, and time distribution of flow.
- Assessment of the tides and currents that are anticipated within the region.
- Assessment of wind effects and other geophysical phenomena that may contribute to specific influences on the area of study.
- An identification of the sources which contributes to sedimentation and of sediment profile.
- Identification of sources and the expected quantities and composition of industrial and municipal effluents, non-point contaminants, and tributary constituent concentrations.
- Identification of the aquatic biota of the region and the physical-chemical, and biological factors which influence its behavior.
- Identification of the available hydrographic and other geometric data pertinent to preparation of the model.

These preliminary assessments have the purpose of ensuring the pertinent and availability of data to provide a basis for the selection of the models needed and to provide a basis for planning field sampling and data acquisition programs. The most satisfactory procedure is to plan the numerical modeling and field data acquisition program together. If possible, the basic hydrodynamic model should be operational during the period in which field data are being obtained. One major reason for concurrent model simulation and data acquisition is that anomalies in field data frequently occur, and the numerical model may be used to identify and resolve them.

Water Quality Models Classification

Water quality models are usually classified according to model complexity, type of water body or water resource, and the water quality parameters (dissolved oxygen, nutrients, etc.) that the model can predict. The more complex the model is, the more difficult and expensive will be its application to a given situation. Model complexity could be a function of the following four factors;

• *The number and type of water quality indicators*
In general, the more indicators that are included, the more complex the model will be. In addition, some indicators are more complicated to predict than others; See Table (3-1).

• *The level of spatial detail*
As the number of pollution sources and water quality monitoring points increase, so do the data required and the size of the model.

• *The level of temporal detail*
It is much easier to predict long-term static averages than short term dynamic changes in water quality. Point estimates of water quality parameters are usually simpler than stochastic predictions of the probability distributions of those parameters.

• *The complexity of the water body under analysis*
Small lakes that "mix" completely are less complex than moderate-size rivers, which are less complex than large rivers, which are less complex than large lakes, estuaries, and coastal zones.

Table (3-1): Indicators for selection of water quality models.
(source: world bank, 1998)

Criterion	Comment
Single-plant or regional focus	Simpler models can usually be used for single-plant "marginal" effects. More complex models are needed for regional analyses.
Static or dynamic	Static (constant) or time-varying outputs.
Stochastic or deterministic	Stochastic models present outputs as probability distributions; deterministic models are point-estimates.
Type of receiving water (river, lake, or estuary)	Small lakes and rivers are usually easier to model. Large lakes, estuaries, and large river systems are more complex.
Water quality parameters	
Dissolved oxygen	Usually decreases as discharge increases. Used as a water quality indicator in most water quality models.
Biochemical oxygen demand (BOD)	A measure of oxygen-reducing potential for waterborne discharges. Used in most water quality models.
Temperature	Often increased by discharges, especially from electric power plants. Relatively easy to model.
Ammonia nitrogen	Reduces dissolved oxygen concentrations and adds nitrate to water. Can be predicted by most water quality models.
Algal concentration	Increases with pollution, especially nitrates and phosphates. Predicted by moderately complex models.
Coliform bacteria	An indicator of contamination from sewage and animal waste
Nitrates	A nutrient for algal growth and a health hazard at very high concentrations in drinking water. Predicted by moderately complex models.
Phosphates	Nutrient for algal growth. Predicted by moderately complex models.
Toxic organic compounds	A wide variety of organic (carbon-based) compounds can affect aquatic life and may be directly hazardous to humans. Usually very difficult to model.
Heavy metals	Substances containing lead, mercury, cadmium, and other metals can cause both ecological and human health problems. Difficult to model in detail.

The level of detail required from modelling can vary tremendously across different management applications. At one extreme, there might be an interest in the long-term impact of a small industrial plant on dissolved oxygen in a small, well-mixed lake. This type of problem can be addressed with a simple spreadsheet and solved by a single analyst in a month or less. At the other extreme, as an example, if there is a need to know the rate of change in heavy metal concentrations in the Mediterranean Sea that can be expected from industrial and agricultural practices in the Nile Delta, the task will probably require many person-years of effort with extremely complex models and may cost a lot of money.

For indicators of aerobic status, such as biochemical oxygen demand (BOD), dissolved oxygen, and temperature, simple, well-established models can be used to predict long-term average changes in rivers, streams, and moderate-size lakes. The behaviour of these models is well understood and has been studied more intensively than have other parameters. Basic nutrient indicators such as ammonia, nitrate, and phosphate concentrations can also be predicted reasonably accurately, at least for simpler water bodies such as rivers and moderate-size lakes. Predicting algae concentrations accurately is somewhat more difficult but is commonly done in the United States and Europe, where eutrophication has become a concern in the past two decades. Toxic organic compounds and heavy metals are much more problematic. Although some of the models reviewed below do include these materials, their behaviour in the environment is still an area of active research. Models can cover only a limited number of pollutants. In selecting parameters for the model, care should be taken to choose pollutants that are of concern and are also representative of the broader set of substances which cannot all be modelled in detail.

Model Type and Solution Techniques

According to Dahl and Wilson 1997, when models are classified by the modelled ecosystem, distinctions are made between lake models, river models etc. Models can, however, be categorized by other criteria as mentioned earlier. The categorization of models according to their mathematical description and the (solution) techniques used to solve them is seen in Figure (3-2), this classification distinguishes between steady state and dynamic models, black box and fundamental models, continuous and discrete time models and between deterministic and stochastic models. Furthermore models are classified according to the dimension of the model (1D, 2D or 3D). For most problems many kinds of models can be used. In general, the simplest model structure that is able to solve the problem is chosen. The choice of model structure is further determined by presence of data, technical and personal resources, computer capacity etc.

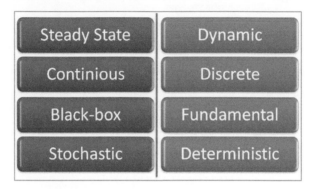

Figure (3-2): Categorization of models according to their mathematical description and the (solution) techniques (After Dahl and Wilson, 2000)

Classification of Models According to Mathematical Description:

1. Steady State / Dynamic Models
A steady-state model involves only algebraic equations as opposed to dynamics models which, incorporating derivatives, can model evolving conditions with time and/or space. Statistical regression models which describe only the averaged status of the lake given constant inputs and external loading forcing functions are typical steady-state models. For applications steady-state models are appropriate where either the dynamics are very fast or very slow compared with the time scale of interest. In the case of fast dynamics, equilibrium values can be used in the model and in the case of slow dynamics, initial conditions can be employed with little loss of accuracy.

2. Input – Output / Physically Based Models
In a black box or data-driven model, both structure and parameter values are determined by the data and any prior knowledge about the system is not utilized. A characteristic feature of black box models is that they require large amounts of data to get good results. On the other hand, traditional modelling of physical processes is often named *physically-based modelling or (knowledge –driven modelling)* because it tries to explain the underlying processes. *In a physically based model* the structure of the system, developed from known scientific laws, is the base for the model. The structure is normally built up by a number of subsystems and the interactions between them. Unknown parameters in the model are determined by statistical methods to get good agreement between model results and actual data. As opposed to a black box model, the parameters in a physically based model have a physical meaning. A physically based model can be used with more confidence for extrapolation outside the area of observations than an empirical model. As an example of such models is a hydrodynamic model based on Navier-Stocks partial differential equations numerically solved using finite difference scheme.

3. Continuous Time / Discrete Time Models

A continuous time model is specified by a set of differential equations. As conservation equations are best described as differential equations, this means continuous time models are a natural way to describe much of the important phenomena in aquatic modelling. An alternative description is difference equations that are better suited for applications where samples are taken at regular intervals, or indeed if absolute time is unimportant, but only a sequence of events. Water quality data are discrete samples, and therefore suitable for difference equations, (Orlob 1983), although this does not preclude the use of continuous models. The disadvantage of discrete time models is that while the parameters in continuous time models are directly related to the well understood conservation equations, the equivalent discrete time model parameters are non-trivially related, carrying a mixture of sampling time information and other parameters. This makes the statistics considerably more complex.

4. Deterministic / Stochastic Models

Another classification arises between models that contain uncertainty or randomness in their final results and those that do not. Stochastic models are models in which the final outcome is not known with precision by the present state and the future values of external variables (inputs) of the model. Stochastic models also take into account the random influences of the temporal evolution of the system itself. Although the stochastic description of, for example, environmental systems may be more realistic, the large majority of environmental models formulated so far are deterministic. The main reason for this fact may be the lack of data for the characterization of random variables, high requirements of computational resources for solving stochastic differential equations and the success of deterministic models in describing average future behavior, (Jeppsson, 1996). Every deterministic water quality model must be based on the following general principles: conservation of mass, momentum and energy; narrow bands of biomass composition; boundary conditions and initial conditions; laws governing chemical, biochemical, and biological processes; and the second law of thermodynamics." (Mauersberger 1983).

Classification of Models according to dimensions

Water quality models are naturally spatially distributed and can have up to four independent variables: three spatial dimensions (x, y, and z) and time. Normally the x-dimension is parallel to the flow, the y-dimension is right angles to the flow and the z-dimension is depth. In larger lakes x-direction can be north-south and y-direction east-west or similar. To handle such a model computer tools are necessary. See for example Cerco and Cole (1993).

Models with a single independent variable (usually time) are described with ordinary differential equations (ODE) while models with more than one independent variable (time and a spatial distribution) are described with partial differential equations (PDE). In every reduction of the number of independent variables, the model loses ability to predict reality, But the model gets much easier to construct, verify and solve with fewer independent variables. The modeller must find a model structure that is simple enough to be possible to construct and calibrate and still not to be over simplified so that its predictive capacity is lost.

Time Dimension

In the counting of dimensions for water quality modelling purposes, the time dimension is not normally included. A dynamic lumped parameter (zero dimensional) model still retains the time dimension, but is considered a zero dimensional model. A full 3D-model has the three physical dimensions (x, y and z dimension), and is considered a 3D model both if it is steady-state (no time dimension) or dynamic (with time dimension).

Zero spatial dimensions

Lumped parameter or zero-D models consider the lake as a single well-mixed container. Lumped parameter models are used to model complex chemical kinetics or dynamics of biological systems. Phenomena like temperature and flow of water are often modelled using distributed models. Lumped parameter models are also used for long time studies; where as distributed models are used for short term transport phenomena.

One spatial variable (1D)

1D model are commonly used in two different situations; models of rivers and deep and narrow lakes. The justification of the 1D approach has been argued as "It may be questioned to what extent a one-dimensional model adequately represents a natural system such as a lake or reservoir. From the point of view of the hydrodynamics, provided that the lake is neither extremely long and narrow, nor extremely broad and shallow, a one dimensional hydrodynamic approach is satisfactory." Orlob (1983a) (Hamilton and SCHL-adow 1997,) see some shortcomings of the 1D modelling approach: "By contrast, water quality variables, for example nutrient concentrations, exert a neglible effect on the density distribution, and therefore could potentially display a two- or three-dimensional distribution despite a 1D density distribution." This is a reason for using a 3D model, but a little later in the same text, Hamilton and SCHL-adow (1997) argues for the usage of a 1D model anyway. The reasons are that a 3D-model is harder to construct, harder to verify and its harder to know all the initial conditions. Therefore a 1D model can give greater certainty in the results. "While this is recognized as a shortcoming of a 1D model, it does not necessarily imply that a multi-dimensional approach would produce a more correct picture." It's hard to know all the initial conditions for a 3D-system and it's hard to verify the results. Therefore a 1D model with greater certainty in the results could be preferred

Two spatial variables (2D)

Shallow waters, like estuaries, where the water quality is homogenous in depth-direction, are modelled using depth averaged 2D models. For reservoirs there exist 2D vertical-longitudinal hydrodynamic models.

Three spatial variables (3D)

The 3D model is the ultimate aim of water quality models since it is capable of modelling flow and temperature profiles in large lakes, where both horizontal and vertical movements are of significance for the water quality. The two competing numerical techniques for this class of problems are finite elements (or the closely related finite difference), and approximating the distributed nature with a series of interconnected well-mixed volumes modelled as a large and sparse collection of ordinary differential equations.

Compartment models

A common simplification when modelling a distributed system is to view the system as a set of fully stirred tanks, connected to each other so that the outflow from one tank is the inflow of another. Each tank, or compartment, contains an ODE, and the models are called compartment or compartmental models. Compartment models make it possible to model distributed phenomena with ODEs, and avoid PDEs.

3.3.3. Surface Water Quality Modelling Requirements

Modelling is one of the main tools in understanding the surface water problems and finding appropriate solutions to it. Today, surface water quality management has moved well beyond the urban point-source problem to encompass many other types of pollution. In addition to wastewater, we now deal with other point sources such as industrial wastes as well as non-point sources such as agricultural runoff (Chapra, 1997). Over the past 75 years, engineers have developed water quality models to simulate a wide variety of pollutants in a broad range of receiving waters. In recent years, these receiving water models are being coupled with models of watersheds, groundwater, bottom sediments, to provide comprehensive frameworks predicting the impact of human activities on water quality. As Thomann (1998) terms it, a "Golden Age" of water quality modelling is upon us (Chapra, 2003).

For years, the analysis of water quality has concentrated on the dissolved oxygen (DO) and biochemical oxygen demand (BOD). The balance between DO and BOD concentrations was the result of two processes: the reaeration of the water column, and the consumption of DO in oxidation of BOD. Later emphasis has been on extending and refining the Streeter-Phelps formulation by using a more generalized mass balance approach and by the inclusion of additional processes such as benthic oxygen demand, benthic scour and deposition, photosynthesis and respiration of aquatic plants, and nitrification. The more comprehensive water quality models have been developed to include the nitrogen and phosphorus cycle and the lower trophic levels of phytoplankton and zooplankton. A number of investigations have modeled the algal nutrient, Silica.

A number of chemical constituents have been selected and modeled by assuming thermo-dynamic equilibrium. The fate of toxicants such as pesticides, metals, and polyCHL-aorinated biphenyls (PCBs) is very complicated involving adsorption-desorption reactions, flocculation, precipitation, sedimentation, and biological uptake. Examination of toxicants and their impact on biological populations requires ecological models. The water quality modeling methodology requires considering the following:

- The water quality parameters to be modeled
- The dimensional and temporal resolution of the model
- Data requirements for model building

Water Quality parameters

The water quality parameters most frequently simulated include salinity, light, temperature, DO, BOD, coliform bacteria, algae, nitrogen, and phosphorus . Each of these parameters interacts with each others, but the significance of their dependencies varies among constituents, and their inclusion in a numerical water quality model depends upon the study objectives and the water body under consideration.

Dimensional and Temporal Resolution of Model

In a numerical water quality model the choice is between a 1-dimensional model and one that incorporates two or three spatial dimensions. A long, narrow, and vertically well-mixed water body may be represented by a one-dimensional model consisting of a series of segments averaged over the cross section. Where there is pronounced vertical stratification, it is likely that a laterally averaged two-dimensional model will be needed. In other situations where there are marked lateral heterogeneities in water quality but the water body is well mixed, a vertically averaged two-dimensional model is indicated. If significant lateral heterogeneities are accompanied by pronounced stratification, a three-dimensional model may be required. Most existing water quality models are one-dimensional. Practical applications of two-dimensional depth and breadth integrated models have been made and are feasible. Three-dimensional water quality models are presently research tools; data requirements for calibration and verification make them prohibitively expensive at present for practical application.

The basis of all water quality models is a velocity field either specified by empirical measurements or computed by numerical hydrodynamic models. The current trend in hydrodynamic modeling is toward development of 3-dimensional models with increased spatial and temporal resolution in order to resolve important scales and to minimize the need for parameterization. As a result, modern time-dependent hydrodynamic models normally have time steps on the order of minutes to one hour. The chemical and biological equations of water quality models have characteristic time scales determined by the kinetic rate coefficients. These time scales are usually on the order of one to ten days. The phenomena of interest, such as depletion of DO and excessive plant growth, occur on time scales of days to several months.

Direct coupling of hydrodynamic and water quality models provides potential spatial and temporal resolution that cannot be effectively interpreted. The reasons are that present field sampling programs resolve constituent concentrations on the order of a kilometer to tens of kilometers in the horizontal, meters in the vertical, and days to weeks in time. In addition, the kinetic rate coefficients presently used in water quality models resolve dynamics on the order of days to weeks. Direct coupling substantially increases the cost of computation. Direct coupling is necessary only for those constituents such as temperature, salinity, and suspended particulates, whose contribution to density gradients may substantially affect the flow. Uncoupling permits averaging of the hydrodynamic model output, which results in less costly water quality computations. There are several conducted studies and research in the field of water quality modelling. For both streams and lakes the following section gives an idea of the progress in the field of water quality modelling.

3.3.4. Water Quality Modelling of Streams and Catchments

Many of the stream models in use are extensions of two simple equations proposed by Streeter and Phelps in 1925 for predicting the biochemical oxygen demand (BOD) of various biodegradable constituents, and the resulting dissolved oxygen concentration (DO) in rivers (Thomann, 1972). Often used with these BOD-DO models are other fairly simple first order exponential decay, dilution, and sedimentation models for additional non-conservative and conservative substances (Mauersberger, 1983). Various modifications of Streeter and Phelps model have been proposed to take into account other water quality processes that take place in natural streams. Some examples on developed models are highlighted here. Several models have been developed since then, such as DOSAGE1, which is a stream quality model developed by Texas Water Development Board (1970a) to simulate the spatial and temporal variations in BOD and DO under various conditions of flow and temperature.

The other famous model is QUAL1, which was developed also by Texas Water Development Board (1970b). It was designed with capabilities to simulate the spatial and temporal variations in the water temperature and conservative mineral concentration in addition to BOD and DO. Another good example of many of the operational aquatic ecosystems simulation models is QUAL2, developed by the US Environmental Protection Agency (Water Resources Engineers, Inc., 1973). This model predicts the effects of various constituents, especially nutrients on the aquatic ecosystem. Since the 1970's a variety of models have been developed and applied to simulate variations of water quality parameters whether from point or non-point sources and whether in steady state or time varying conditions.

Many models have been applied and many studies have been conducted on various rivers and streams, on catchment and watershed scale. From the recent applications, Radwan (2002) applied MIKE 11 modelling package to study the river water quality of the Molenbeek brook and the Mark brook sub-catchments of the river Dender, Belgium. The concentration of BOD, DO, ammonia and nitrate were calculated using MIKE 11 developed by Danish Hydraulics Institute as a detailed modelling system for simulation of flows, sediment transport and water quality in rivers, estuaries and other water bodies.

In 2004, Lopes et al., developed a hydrodynamic and water quality model for a regulated rive segment. The objective of the model was present a global study on the hydrodynamics, water quality and their influence on aquatic fauna. The case study was conducted on a segment of the Lima river (North Portugal), downstream of the Touvedo dam, which was mainly constructed for hydroelectric power production. The ISIS FLOW program was used to simulate the hydrodynamics. This model also generates the necessary input data for the water quality simulation using the ISIS QUALITY module. Two basic principles govern the ISIS programs: conservation of mass and momentum. Water quality parameters (dissolved oxygen and temperature) were simulated for different operational conditions of the Touvedo dam: different discharges (water quantity and duration) and two levels of the water withdrawal, where different water characteristics have been measured.

A three-dimensional water quality model (CCHE3D WQ) was developed by (Chao et al., 2007) for simulating temporal and spatial variations of water quality with respect to phytoplankton, nutrients, and dissolved oxygen. Four major interacting systems were simulated, including phytoplankton dynamics, nitrogen cycle, phosphorus cycle, and dissolved oxygen balance. The effects of suspended and bed sediment on the water quality processes were also considered. The model was verified using analytical solutions for the transport of non-conservative substances in open channel flow, and then calibrated and validated by applying it to the study of the water quality of a natural shallow oxbow lake. The simulated time serial concentration of phytoplankton (as CHL-aorophyll) and nutrients were generally in good agreement with field observations. Sensitivity studies were then conducted to demonstrate the impacts on water quality due to varying nutrients and suspended sediment loads to the CHL-aorophyll concentration

3.3.5. Water Quality Modelling of Lakes

Mathematical modelling of lakes water quality started to receive high attention in the 1960's. According to Jorgensen (1983), mathematical models of lakes have evolved along two different lines. First there was the extension of the zero-dimensional model to one-, two-, and three-dimensional models. Then there were the modelling activities that focussed primarily on a better and more detailed description of the chemical-biological processes (Park *et al.,* 1974; Chen and Orlob, 1975; Jorgensen, 1976; and others). From the survey of the literature many lake models have been applied in various regions, and as a result of several applications, models have become more and more complex.

Modelling of water quality in lakes involves the representations of effluent quality, mixing pattern, physical and chemical processes and biological growths and their role in the removal and release of substances. Such models can be classified into physical, chemical or biological models that simulates lakes europhication. Another classification may be as long term planning models or short term operational models. Several water quality studies have been performed on lakes in general and on shallow lakes in particular. It is important to understand the physics of hydrodynamics of shallow lakes, before water quality modelling can be used to manage lakes water quality. The hydrodynamic behaviour in shallow lakes is governed by several factors that are explained below.

Characteristics of Hydrodynamic Modelling in Shallow Lakes

Physical processes in coastal lagoons or shallow lakes are influenced most by winds, tides and morphometry. Among the most important morphometric factors are: sea inlet dimensions; width to length and depth ratios of the shallow lake or lagoon; bottom topography; and, mean depth. Inlet dimensions control the exchange of water, including the dissolved and suspended material that it contains, and this in turn determines flushing rates and residence times. Bottom topography, including both natural and man-made channels, plays an important role in guiding the tidal and non tidal circulation of shallow lakes. Mean depth may be the most important of the three geometric factors. Lagoons or shallow lakes are characteristically shallow, with a large horizontal to mean depth ratio, and several hydraulic and hydrographic features arise as a direct consequence.

First, shallow water is especially responsive to heating and cooling processes. Even over diurnal time scales, warming and cooling is distinct. Second, wind effects cause intense vertical mixing and significant wave action and these mixing processes can extend to the bottom. As a result, shallow lakes or lagoons favor vertical homogeneity, and significant density gradients in the vertical are uncommon. Third, in shallow water the bottom frictional influence may extend to the surface, and currents are quickly damped once forcing ceases. To maintain well-defined circulation patterns forcing must be continuous. Tidal forcing is continuous in a periodic sense, but tidal motions arise from exchanges with adjacent continental shelf waters, and inlet jetting can result in rapid dispersion and attenuation of the tidal current. Thus, as noted above, inlets may have a significant constricting effect.

Tidal motions may be significant only in the immediate vicinity of the inlet. Wind forcing can be both local and non-local (by changing sea level outside the inlet), but it is intermittent in time, and variable in speed and direction. Thus, near inlets, tidal currents are often dominant, while in the interior of the lagoon, wind forcing is primarily responsible for maintaining the circulation of the lagoon. Finally, shallow water enhances the residual or net motion that occurs over any complete tidal cycle. The circulation of a coastal lagoon or shallow lake is highly variable in both space and time.

The following section illustrates a selection of research progress in water quality and ecological modelling in lakes. These studies shows the application of existing water quality modelling tools and the development of specific tailored models depending on the objectives of water quality modelling and the type of lake and associated ecological system.

Collins (1988) performed a study on lake water quality evaluation for management. A set of lake-evaluation techniques that can be used as early indicators of anthropogenic stress on the aquatic ecosystem were designed. These approaches were used to predict the response of the ecosystem to changes in land- and water-use practices in an effort to protect or restore water quality. In this study the mathematical model CE-QUAL-R1 was used with lake water-quality monitoring, and algae nutrient bioassay experiments to assess the trophic status of Lake George, New York and to test suitable management strategies.

Another study was conducted by Stefan et al. (1993) on model simulations of dissolved oxygen characteristics of Minnesota Lakes. A deterministic, one-dimensional, unsteady numerical model has been developed, tested, and applied to simulate mean daily dissolved oxygen (DO) characteristics in 27 lake classes in the state of Minnesota, USA. Results of the study showed changes of DO due to climate change effects. Such changes would alter water quality dynamics in lakes and have a profound effect on lake ecosystems including indigenous fishes. The results presented are useful for evaluating environmental management options.

A lake ecological model was developed by Sagehanshi et al., 2001. In this study, an ecological model describing the ecosystem of the Keszthely Basin, Lake Balaton, Hungary, one of the typical shallow and eutrophic lakes, was proposed. This model includes three types of zooplankton and two types of fish as well as two types of algae and nutrients. Parameters concerning the algae and fish were estimated based on observations in the basin between 1991 and 1995. The other parameters and the structure of the model were determined in an earlier study. The parameters of the model were calibrated with the Monte Carlo technique, and its predictability was confirmed. The effects on the basin's ecosystem of three restorative manipulations, namely a biomanipulation, reduction of loading phosphorus, and dredging the sediment, were assessed by simulation studies using the proposed model.

Del Buttcher (2003) conducted a study on approaches for nutrient management in the lake Okeechobee Watershed, USA. Phosphorus has been identified as the limiting nutrient for algae growth, and therefore has been the focus of in-lake and watershed restoration programmes. Watershed modeling is playing a critical role in assisting water resource planners and regulators in spatially and temporally quantifying the existing conditions throughout the Okeechobee watershed, as well as providing cost-effective information for various Best Management Practices scenarios. The Watershed Assessment Model (WAM) was selected because of its GIS-based structure that provides water and nutrient flows throughout the entire stream network. This study indicates the importance of integrating new tools with mathematical models such as GIS for better understanding, visualization and analysis of the modeled system.

The difference in catchments and hydrology between lakes leads to a difference in nutrient loading and water residence time. Management options to reduce the trophic state have to be analysed. (Van Puijenbroek et al., 2004), presented the LakeLoad model which calculates the load of nutrients for 41 polder lakes in The Netherlands. Detailed hydrological information on the complex water system of the polders is used to determine the lake catchment. A lake may loaded as a result of run-off and leaching in the catchment, atmospheric deposition, point-source emissions and the inlet of water from outside the polder. These input fluxes are modelled separately and the input is retrieved from other models and databases. The output of this model is input to the ecological model, PCLake, which calculates the growth of algae, fish and plants in the lake. Using these models it is possible to calculate the effects of differences in agricultural practice or the reduction in point sources on the ecological state of the lake. The effect of water management options, such as phosphorus removal from inlet water, or reducing inlet water by allowing flexible water levels, can also be modelled. Dahl et al., 2006, developed a combined suspended particle and phosphorus water quality model was developed and applied on Lake Vänern in Sweden . The model was modified to handle two separate sub-basins, but increasing the horizontal resolution further by splitting the basins into coastal area and pelagial failed, as the model fit to experimental data deteriorated. Besides, the scant reference data available for the coastal areas makes this a dubious exercise. Parameters for the nutrient dynamics in the water column required less tuning (up to 60%) than the sedimentation and sediments (up to a factor 70). The fit to experimental data is good for the periods between 1900 and 1940 and that after 1980, but is less satisfactory for the more polluted conditions in the middle of the century. The model is applied to two scenarios: increased emissions from a pulp and paper mill by the lake, and decreased phosphorus emissions achieved by a combination of effects on farmland, woodland, and rural households. These two scenarios demonstrate the usefulness of a dynamic quantitative lake water quality model.

Zacharias and Gianni, (2008) studied the water circulation inside the re-flooded Drana lagoon using a hydrodynamic model. Several cases were investigated based on different driving forces, such as tide, wind and fresh water inflows. And also the study dealt with an advection/dispersion model that was used for the spatial and temporal simulation of both temperature and salinity. Oceanographical and meteorological field measurements were used for the calibration and validation of both models. The computed daily water temperature variations were well correlated with the data gained during the field observations.

3.4. APPLICATIONS OF REMOTE SENSING IN SURFACE WATER QUALITY MODELLING

3.4.1. Overview of Data Supporting Water Quality Remote Sensing

The traditional measurement of water quality requires in situ sampling, which is a costly and time-consuming effort. Because of these limitations, it is impractical to cover the whole water body or obtain frequent repeat sampling at a site. This difficulty in achieving successive water quality sampling becomes a barrier to water quality monitoring and forecasting (Senay et al. 2001). It would be advantageous to watershed managers to be able to detect, maintain and improve water quality conditions at multiple river and lake sites without being dependent on field measurements (Shafique el al., 2002). Remote sensing techniques has the potential to overcome these limitations by providing an alternative means of studying and monitoring specific water quality parameters such as total suspended matter and CHL-aorophylla over a wide range of both temporal and spatial scales.

 Several studies have confirmed that remote sensing can meet the demand for the large sample sizes required for water quality studies conducted on the watershed scale (Senay el al., 2001). Hence, it is not surprising that a significant amount of research has been conducted to develop remote sensing methods and indices that can aid in obtaining reliable estimates of these important hydrological variables. These methods ranged from semi-empirical techniques to analytical methods for estimating and producing quantitative water quality maps. Several researchers have developed regression formulas to predict several lake water quality parameters from satellite data by employing spectral ratios or indices. These water quality parameters have included CHL-aorophyll a concentration, suspended matter concentration and turbidity.

Lately, a few studies have used remote sensing data collected from different platforms, such as ground-based, airborne, and satellite for mapping and monitoring of water quality. This section reviews the literature particularly how ground based; airborne and satellite-based remote sensing data were used in the mapping and monitoring of water quality.

Ground-Based Data Collection

Traditionally, ground-based collection of water-quality data is done using field samples or field spectrometers. Field samples are collected by hand at many geographically specific locations in the water body and taken back to a laboratory for analysis. Lake temperature, pH, dissolved oxygen are taken in the field using instruments designed for that purpose. The most useful measurements taken by a field spectrometer are those coordinated with acquisition of aircraft or satellite data (Campbell, 1996). This is a necessary step in aerial or satellite-based remote sensing because images obtained from remote sensors are not immediately comparable to ground truth data or field spectrometer data because of atmospheric distortion (Aspinall, Marcus, & Boardman, 2002). Nordheim and Lillesand (2004) measured upwelling and incoming radiation over 2048 spectral bands from 339.99 nanometers (nm) to 1023.9 nm for the purpose of developing a CHL-aorophyll *a* estimation algorithm. Using two spectrometers, above and below the surface of lakes in Northern Wisconsin and Michigan, reflectance spectra were taken at 6 sites directly above the surface and 6 sites immediately below the surface. They then classified the lakes into classes based on absorption spectra, oligotrophic or blue lakes, mesotrophic or green lakes, and high dissolved organic carbon (DOC) or black lakes. They concluded that their estimates of CHL-aorophyll concentration in "blue" and "green" lakes were reasonable, while their estimates in "black" lakes were questionable. From the available literature, it is clear that the use of *in situ* field samples is an extremely accurate method of determining water quality and measuring water-quality variables. However, this method gives a limited spatial representation of the study area as well as being costly and time consuming (Ritchie & Cooper, 2002). Using a spectrometer, water samples, and regression analysis, Rao (1998) was able to estimate CHL-aorophyll *a* concentration with great accuracy. Her estimates of turbidity and suspended sediments were also very good. It appears that the use of spectrometer readings, water samples, and regression analysis is a good way to determine CHL-aorophyll *a* concentration, turbidity, and suspended sediments based on the spectral signatures of the water body.

Airborne Remote Sensing Data

Airborne remote sensing data is gathered by a number of different sensors, Multi Spectral Scaners (MSS), which collect data in a few bands, and hyperspectral sensors, which collect data in a large number of spectral bands. Monitoring of water-quality parameters, such as CHL-aorophyll *a*, turbidity, and suspended sediments, is done more effectively with hyperspectral data because there are more bands and better spectral resolution. Higher spectral resolution is required to measure the fine features of water quality (Ritchie & Cooper 2002). Hyperspectral remote sensing of inland water bodies, like lakes and rivers, can be a useful tool for monitoring water-quality parameters according to Shafique et al., (2001). Airborne hyperspectral remote sensing, can be used to measure CHL-aorophyll *a*, total suspended solids, turbidity, and Secchi depth, which is a measure of suspended materials. (Dekker, Malthus, & Seyhan 1991; Gittleson, 1992; and Kallio, Kutser, Hannonen, Koponen, Pulliainen, Vepsalainen, & Pyhalahti, 2001). The many bands available in the hyperspectral sensors allow researchers to detect these water quality parameters unlike the few coarse bands of the multispectral sensors. By analyzing the hyperspectral profile of individual pixels or groups of pixels, one can compare those profiles to library spectra, profiles of known concentrations, to determine CHL-aorophyll concentration, suspended sediment, etc. A study on a river system in southwest Ohio that included the Great Miami River, a tributary of the Ohio River, was conducted by Shafique, Autrey, Flotermersch, & Fulk (1999).

Using the Compact Airborne Spectrometer Imager (CASI). Using ENVI software, Shafique et al., (2002) determined that the two bands, 672 nm and 705 nm, best correlated the ground truth data with the remote sensing data. They stated that the use of hyperspectral imagery to assess water quality issues of CHL-aorophyll *a* and turbidity was an accurate way of mapping spatially continuous data. Karaska, Huguenin, Beacham, Wang, Jensen, and Kaufmann (1999) studied the Neuse River of North Carolina using Airborne Visible/InfraRed Imaging Spectrometer (AVIRIS) hyperspectral images to measure CHL-aorophyll, suspended minerals, DOC, and turbidity to determine source points of pollution. The team concluded that the hyperspectral sensor AVIRIS has the ability to directly measure CHL-aorophyll concentrations in the Neuse River. In the case of dissolved organic carbon and suspended minerals, field sampling was inadequate to assess these values. Therefore the team concluded that the accuracy of hyperspectral measurements of suspended minerals, dissolved organic carbon, and turbidity measurements remains uncertain.

It is evident from the literature review that the use of airborne remote sensing of water quality is a superior method of monitoring water quality of inland water bodies due to the speed, accuracy, and amount of data that can be collected. The use of airborne multispectral sensors for water quality monitoring is limited by the small number of spectral bands available with this type of sensor and the fact that many bands are required to monitor the large number of water quality parameters involved in water quality (Ritchie & Schiebe 2000). Therefore, hyperspectral sensors are optimal for this task. It has been demonstrated that CHL-aorophyll, turbidity, total suspended solids, and Secchi depth can be determined by this method. (Dekker,1991).

3.4.2. Integrating GIS, Remote Sensing and Modelling for Surface Water Quality Management

A GIS processes any data that has a spatial component. Remotely sensed data can be best utilized if they are incorporated in a GIS that is designed to accept large volumes of spatial data. Beside data handling and processing, specific applications of of GIS and remote sensing have mainly concentrated on non-point source pollution (NPS). This is because remotely sensed data products such as land-use/land cover could be directly utilised in NPS modelling. Several watershed models have been interfaced with GIS including the export coefficient model, AGNPS and DRASTIC. Agricultural Non Point Source Pollution (AGNPS) model estimates nitrogen, phosphorous and chemical oxygen demand concentration in runoff and assesses agricultural impact on surface water quality based on spatially varying controlling parameters (e.g. topography, soils, land-use, etc...). Kim and Ventura (1993) managed and manipulated land-use data for modeling NPS pollution of an urban basin using an empirical urban water quality model. Another approach uses an export coefficient model to calculate nutrient losses from catchment to surface water mainly in terms of aerial extent of different land-use/land cover and their associated fertilizer application rates. Matikalli et al. (1996) derived historical land cover data from air-born and satellite sensors, and implemented the model using Arc/Info to estimate historical nitrogen and phosphorous loading in river Glen watershed in the UK.

As important as the improvements in the quality and availability of remote sensing data is the growing number of geospatial data management and analysis tools available for use in different application of water resources. With geographic information systems (GIS), digital remote sensing data can now be integrated with other types of digital data and with models. Such technological advances can be useful for development of remote sensing applications for water resources and water quality management. However, the use of remote sensing data and applications involves more than the underlying technical capacity. From the perspective of the remote sensing applications end users, what is important is the information that remote sensing applications can make available, not the raw data that it can provide. Equally important, is the ability and positive acceptance of applying the recent technological advances. This depends on institutional, leadership, budgetary, procedural, and even personnel factors.

Applications of Remote Sensing in Surface Water Quality

Suspended sediments
Suspended sediments could be a common pollutant both in weight and volume in surface waters of freshwater systems (Lal, 1994). Suspended sediments increase the reflected energy from surface waters in the visible and near infrared proportion of the electromagnetic spectrum (Ritchei et al., 1976) Figure (3-3). *In situ* and controlled laboratory measurements have shown that surface water radiance is affected by sediment type, texture and color (Holyer 1978; Novo et al., 1989a; Han and Rundquist 1996), sensor view and sun angles (Ritchei et al., 1975; Novo et al., 1989b; Ferrier 1995), and water depth (Mantovani and Cabral 1992).

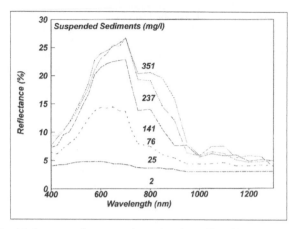

Figure (3-3): The relationship between reflectance and wavelength as affected by the concentration of suspended sediment (Ritchei et al., 1976).

Since the mid 1980's remote sensing studies of suspended sediments have been made using data from satellite platforms such as Landsat (Kritikos et al. 1974; Carpenter and Carpenter 1983; Khorram 1985; Ritchie et al. 1990; Harrington et al. 1992), SPOT (Lathrop and Lillesand 1989; Froidefond et al. 1993), IRS (Chubey and Subramanian 1992), AVHRR (Strumpf and Pennock 1991; Froidefond et al. 1993), and CZCS (Coastal Zone Color Scanner) (Amos and Toplis 1985; Mayo et al. 1993). These studies have shown significant relationships between suspended sediments and radiance or reflectance from spectral wave bands or combinations of wave bands on satellite and aircraft sensors.

Many studies have developed statistical relationships (algorithms) between concentration of suspended sediments and the radiance or reflectance for a specific date and site. Few studies have taken the next step and used these algorithms to estimate suspended sediments for another time or space (Whitlock et al. 1982, Curran and Novo 1988; Ritchie and Cooper 1988, 1991). Once developed, an algorithm should be applicable until some cathcment event changes the quality (size, color, mineralogy, etc.) of suspended sediments delivered to lake.

CHL-aorophyll-a
Another application of remote sensing is the measurement of CHL-aorophyll-a. Lakes and other water bodies depend upon their catchments for nutrients and other substances to sustain biological activities. While these nutrients and substances are required for a healthy aquatic environment, an excess of these inputs leads to nutrient enrichment and eutrophication of the lake as discussed earlier. The rate of eutrophication depends on topography, soils, land use, and runoff on the contributing catchment. Eutrophication of a water body can be quantified in terms of trophic level or concentration of the CHL-aorophyll contained in the algae/phytoplankton cells. While aging of water bodies is a normal process, better information on eutrophication (CHL-aorophyll content) in a lake will allow better management plans to be developed to control the source in nutrient input from the catchment.

Monitoring the trophic level or concentration of CHL-aorophyll-a (algal/phytoplankton populations) is the key to managing eutrophication in lakes. The algal population is a water quality parameter that if it is too large can be a problem. Algal concentration can be monitored by collecting samples, extracting CHL-aorophyll, and measuring concentrations in the extracts by photometric techniques in the laboratory. Remote sensing has been also used to measure CHL-aorophyll concentrations and patterns. Most remote sensing studies of CHL-aorophyll in water are based on empirical relationships between radiance/reflectance in narrow bands or band ratios and CHL-aorophyll. Thus field data must be collected to calibrate the statistical relationships or to validate the models developed.

Measurements have been made *in situ* (Quibell 1992; Han et al. 1994;Rundquist et al. 1996) and from aircraft (Dekker et al. 1992; Gitelson et al. 1994; Harding et al. 1995), Landsat and SPOT (Carpenter and Carpenter 1983; Lathrop and lillesand 1989; Strumpf and Tyler 1988; Dekker and Peters 1993) and CZCS (Hovis 1981; Gordon et al. 1983), Figure (3-4). These studies have used a variety of algorithms and wavelengths to successfully map CHL-aorophyll concentrations of oceans, estuaries and fresh waters. Harding et al.

While measuring CHL-aorophyll by remote sensing is possible, studies have also shown that the broad wave length spectral data available on current satellites do not permit discrimination between CHL-aorophyll and suspended sediments (Dekker and Peters 1993; Ritchie et al. 1994) due to the dominance of the spectral signal from suspended sediment. These discoveries suggest new approaches for application of airborn and spaceborne sensors to exploit these phenomena to estimate CHL-aorophyll in surface waters under all conditions. New hyperspectral sensors have been launched and data have become available. Data from several satellite sensors (i.e. Sea WiFS, modular Optical Scanner (MOS), Ocean color and Temperature Scanner (OCTS) are now becoming available and hold great promise for measuring biological productivity in aquatic systems.

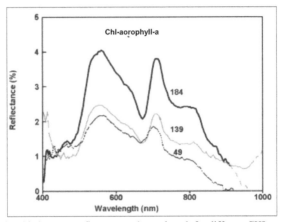

Figure (3-4): The relationship between reflectance and wavelength for different CHL-aorophyll-a concentrations (Ritchei et al., 1976).

Satellites such as (SEA-WIFS, MOS, OCTS, QuickBird, Resource21, Orbview, etc.) and sensors (hyperspectral, high spatial resolution) should provide both the improved spectral and spatial resolution needed to monitor water quality parameters in surface waters in fresh water in lakes and streams, estuaries, and oceans and to differentiate between different water quality parameters. Research needs to focus on the understanding of the relationship between water quality parameters and their effects on the optical and thermal properties of surface waters so that physically based models can be developed (Jerome et al. 1996). Such information should facilitate progress from the empirical approaches now being used to the development of algorithms that will allow the use of the full resolution electromagnetic spectrum to monitor water quality parameters.

In one of the remote sensing applications in water quality, Dekker et al. (2001) applied airborne remote sensing for inland water quality detection and monitoring on Friesland waters, the Netherlands. The study resulted in developing maps of CHL-aorophyll and suspended matter concentrations using analytical algorithms. The study showed that the real problem for getting an emerging technology such as high spectral and radiometric resolution remote sensing accepted and implemented, lies in making it clear to the end-users that the application of the technique is beneficial in their work. Therefore it is necessary to provide the end user with adequate water quality information from remote sensing at the right time in the right format. For this purpose, the study developed a generic methodology in the Netherlands, applicable anywhere in the world, with the following guidelines:

- In operational remote sensing no or few *in situ* measurements should be required, as concurrent sampling and analysis is expensive and labour intensive.
- Standardized operational methods and algorithms for processing of remote sensing data are required to enhance the reproducibility and to speed up the production process. Remote sensing derived information should be available within the same time frame as the laboratory results.
- The development of analytical/deterministic algorithms is a prerequisite in order to develop generally applicable methods with multisite, multisensor and multitemporal validity. Empirical-based approaches always need a large amount of *in situ* data
- Algorithms must offer possibilities for sensitivity analysis and must be suitable to determine the precision of and the errors within results. In this way the end user knows whether or not a particular remote sensing application meets their requirements.

This study also recommended that frequency of availability of remote sensing data is an important issue. Frequency is more easily obtained from Satellite images at lower costs than associated with airborne remote sensing. But a boundary condition is that of sufficiently cloud free images per required monitoring period.

3.5. WATER QUALITY MANAGEMENT INFORMATION SYSTEMS

Before explaining the term "water quality management information systems', the terms *systems, information,* and *management* must briefly be defined. **A system** is a combination or arrangement of parts to form an integrated whole. A system includes an orderly arrangement according to some common principles or rules. A system is a plan or method of doing something. A system is a scientific method of inquiry, that is, observation, the formulation of an idea, the testing of that idea, and the application of the results. The scientific method of problem solving is systems analysis in its broadest sense. Data are facts and figures. However, data have no value until they are compiled into a system and can provide information for decision making.

Information is what is used in the act of informing or the state of being informed. Information includes knowledge acquired by some means. In general, information is the basis for any management and control. Water management activities are not excluded from this general statement. Management measures not based on adequate and reliable information are, principally, unaccountable. There is, therefore, a profound need for effective information that is suitable for such use. As a consequence the development of accountable information systems is receiving much emphasis

Management could be defined as the knowledge of the planning, organizing, directing, and controlling of the resources, so decisions can be made on the basis of facts, and decisions are more accurate and timely as a result. **Management information systems** are those systems that allow managers to make decisions for the successful operation of resources. Management information systems consist of computer resources, people, and procedures. The term **MIS** stands for management information systems. A management information system (MIS) is a system or process that provides the information necessary to manage resources. MIS could be developed for managing any resources, such as human or natural resources, in the case of this research the focus is on Water resources management and specifically water quality management.

In the last decade of this age of information, a shift in awareness of the role of monitoring and information has become apparent. In the past, monitoring originated from the greater scientific ideal that underpins our search for knowledge. The consequence, especially in advanced countries, is that monitoring is frequently, if not implicitly, linked to scientific investigation. Water quality monitoring, world-wide, tends to suffer from a chronic failure to establish meaningful programme objectives. Therefore, it is crucial that when designing management information systems for water quality, it is necessary to have a clear objective of the system uses, the associated tools, and the required monitoring program to support the management plans. In developing countries, responsible organizations for managing water quality acknowledge that they have collected many data, but are unable to answer the basic questions of those related to pollution and water quality issues in a reliable and effective way.

As a consequence, in many countries, data gathering programmes are considered expendable, and are being reduced or even eliminated because there is no clear view of the information product and of the cost-efficiency of monitoring (Ward *et al.*, 1986; Ongley, 1995; Ward, 1995a). In recent years there has been an increasing consensus of opinion that information is meant for action, decision-making and use. Data that do not lead to management action, or for which a use cannot be stated explicitly, are being labelled increasingly as "not needed" (Adriaanse *et al.*, 1995). The next chapter will cover the details of the proposed and developed water quality management information system for agricultural irrigated watershed in data scarce environment.

3.5.1. Information Needs for Management Information Systems Users

Information needs are focused on the three core elements in water management and water pollution control, namely; *the functions and use of water bodies, the actual problems and threats for future functioning*, and *the measures undertaken* (with their intended responses) to benefit the functions and uses (Figure 3-5). As mentioned in the beginning of this chapter, monitoring is the principle activity that meets information needs for water pollution control. Models and decision support systems, which are often used in combination with monitoring, are also useful information tools to support decision making.

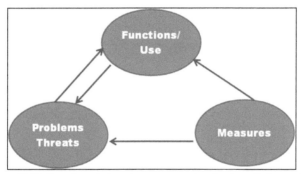

Figure (3-5): Core elements in water management and water pollution control

Once objectives have been set it is important to identify the information that is needed to support the specified objective. The content and level of detail of the information required depends upon the phase of the policy life cycle. In the first phase, research and surveys may identify priority pollution problems and the elements of the ecosystem that are appropriate indicators. Policies will be implemented for these. In the second and third phases, feedback on the effectiveness of the measures taken is obtained by assessing spatial distributions and temporal trends through detailed analysis and application of tools such as modeling. Contaminants may endanger human health by affecting aquatic resources, such as drinking water, and therefore specific monitoring programmes may be initiated to check, on a regular basis, the suitability of such resources. Legislation may also prescribe measurements required for certain decision-making processes, such as the disposal of contaminated dredged material. In the last phase, monitoring may be continued, although with a different design, to verify that control is maintained. The associated information needs change with the respective policy phases (Cofino, 1995). Decision-makers have to decide upon the contents and performance of their desired information products. They are the main users of the information (for management and control action) and they have to account for their activities to the public. Specification of information needs is a challenging task which requires that the decision-making processes of information users are formulated in advance. These different categories of users responsible for water quality management reflect the needs of different stakeholders. Therefore the proposed recommendations of the developed water quality Management Information system (WQMIS) in this study are based on scientific data and models and can account for the stakeholder objectives and needs and focuses on different major watershed components. This is in contrast to traditional decision processes, which have had difficulty in aggregating all of these considerations. The different outputs of such systems can be interpreted in different ways by linking the users to the computer models in a graphical user's friendly interface. This helps to highlight the importance of integrating all sources of water quality data, modelling tools and outputs into one management system.

3.5.2. Decision Support Systems for Surface Water Quality Management

The decision making associated with utilization of water resources is understood here as the process of selecting such actions influencing the behaviour of a given water resources system, which at least intentionally (to make provisions for various false or mislead decisions which have been permanently happening all over the world) should result in a better fulfilment of the goals and objectives by the system under consideration. The decision making can be also understood as process of seeking the "best acceptable" solution for a specific system.

The decision making processes are taking place in a structure consisting of the following elements:
- The **system** (in our case water management system) under consideration representing material and physical reality;
- The **problem** which requires a decision. The term *problem* refers to the existence of a gap between the desired state and the existing state (Sabherwal and Grover 1989). Consequently, the decision making process aims to fill, or at least reduce, this gap and thus solve the problem; and
- The **decision maker**, that is the person or personalized organization, who is required to decide upon the action or a set of actions which are to be undertaken in order to achieve certain objectives (fill or reduce the gap between the existing and desired state of the system). These objectives are provided by those to whom the decision maker is responsible. Most methodologies assume an individual decision maker. However, in a real world situations, the decisions are usually made by a group or even groups of people representing different views, preferences, expectations, etc..

What connects the above mentioned elements of the structure underlying the decision making process is information, which is continuously gathered, exchanged, processed, enhanced, evaluated and used during the decision making processes. Decision making processes associated with water resources management concern many areas and decisions can be purely technical, technical with economic and social impacts, political, economic, social and so on. There is no clear scientific explanation how the decisions are made by individuals, why people make these and not another decisions, what information do they use while making decisions. We can only assume or take for granted, that the decisions can be made faster and better when the decision makers will be provided with the most up-to-date, possibly complete and correct information relevant to decision problem they are confronted with, information relevant to the type of decisions which have to be made.

3.5.3. Definitions and Basic Concepts of DSS

Due to very complex nature of water resources management problems; lack of consistent and complete data; uncertainties and ill-structured form of decision problems, the process of finding decisions cannot be limited to solving of mathematical optimization problems or performing complex simulation. Therefore we will understand the decision support system as a set of computer-based tools that provide decision maker with interactive capabilities to enhance his understanding and information basis about considered decision problem through usage of models and data processing, which in turn allows reaching decisions by combining personal judgement with information provided by these tools.

A decision support system (DSS) could be seen as both a process and a tool for solving problems that are too complex for humans alone, but usually too qualitative for only computers. Multiple objectives can complicate the task of decision-making, especially when the objectives conflict. As a process, a DSS is a systematic method of leading decision-makers and other stakeholders through the task of considering all objectives and then evaluating options to identify a solution that best solves an explicit problem while satisfying as many objectives as possible.

As a tool, a DSS consists of mathematical models, data, and point-and-click interfaces that connect decision-makers directly to the models and data they need to make informed, scientific decisions. A DSS collects, organizes, and processes information, and then translates the results into management plans that are comprehensive and justifiable. Often, water resources stakeholder groups have very diverse goals and values, including environmental, economic, and ecological interests. What complicates this process even further is that water resources managers must try to achieve numerous and often conflicting objectives, such as achieving peak sustainable yield, minimizing environmental impact, managing costs, maintaining adequate water quality, controlling floods, minimizing energy use, and providing recreational opportunities.

DSS programs have been used to develop water resources management plans, adaptable operating rules for water and wastewater systems, and regional policies. Many municipalities and water authorities often derive their water supplies from several sources, which may include surface reservoirs, rivers, groundwater wells or combinations of these sources. To identify the best combination of supply sources in the long term, or to determine the most effective way of managing existing systems, decision-makers need a lot of information to account for all of the hydrologic, hydraulic, water quality, and economic relationships within the system.

Consequently, together with the growing complexity of the decision making problems, there are also growing demands and challenges concerning tools used to provide information and to support decision making processes.

The methodological framework underlying the process of searching for solutions (decisions) of the decision problem is offered by the scientific discipline called *systems analysis* (Sage and Armstrong, 2000), which evolved through parallel developments in mathematics, engineering and economics. As the system analysis has been becoming more mature in recent decades, its applicability in water resources planning and management has been also constantly growing and currently it is impossible to imagine water resources management practice without using methods and tools offered by the systems analysis.

The notion of a *system* is a basic one for this discipline of science (Nandalal and Simonovic, 2002). Physical water resources systems are considered as a collection of various elements interacting in response to natural and human-induced actions. The systems and related human actions are aimed at satisfying social and economic needs. Systems analysis allows to study not only interactions between components of the system, but also to study the overall response of the whole system to various human actions associated with development and/or management alternatives.

The behaviour of the system as a whole or behaviour of some of its components can be subject of systems analysis only then, when the system and/or all its element can be modelled using mathematical representation (mathematical models). Models and their properties can differ very significantly: the same physical phenomenon can be described using different types of models, depending on specific purposes which the model may serve. These different types of models may have different mathematical representation. Therefore the mathematical representations of the reality chosen by the model builder should be consistent with overall accuracy required from the system model and should allow to describe reality adequately to the purpose of the model. This model should provide decision makers with information relevant to decision problem at hand and should address information needs of the stakeholders

A system, understood as a part of physical reality and consisting of a finite number of interrelated and interacting with each other elements, and identified due to functions which this system fulfils, is influenced by uncontrollable and very often not exactly known natural factors on one side and by targeted and aim-oriented human actions on another side. both uncontrollable natural stimuli (uncontrolled inputs) and human-induced, targeted actions (controlled inputs) influence the behaviour of the system, which is "responding" through physical values identified as outputs (system outputs).

Controlled inputs are equivalent to decision variables, which have to be selected in the framework of decision making process from the set of feasible alternatives. The transformation of the system due to influence of both decision variables and uncontrolled inputs is described using a set of so called state variables, which are associated with the mass- and energy preservation. Internal properties of the system are described by system parameters. Finally, physical values through which given system acts at the surrounding are called output variables. The selection of the output variables is very often depending on purpose of the system (or its model).

With the functioning of every system there are also associated certain goals, which should be attained. The functional relationship between decision variables, state variables, system parameters on one side and the quantitative description of the degree to which these goals are attained is called objective function. Depending on complexity of the system and specification of the goals, the objective function may have a form of a scalar (single value) function; but it may also have a form of a vector function attaining multiple values. The process of selecting such values of decision variables, which allow achieving the best possible (with respect to existing constraints on decision and state variables) is called an optimization (Rardin, 1997). If the objective function is a scalar one, we are talking about single objective optimization; when the objective function has a vector representation, the notion of multiobjective (multicriteria) optimization applies (see Rosenthal, 1985, for excellent introduction to this subject or refer to Miettinen, 1998, for extensive presentation of this subject).

As the practice shows, real life decision making problems only very rarely, if at all, boil down to solving clearly-cut optimization problems. The search for solution of the decision problem involves complex patterns of using optimization and simulation models of the system under consideration in order to find feasible and satisfactory values of decision variables (controlled inputs) in a framework of decision making processes. The system model, consisting very often of many sub-models and components, must also encounter for the presence of uncontrolled inputs influencing the system at hand. The information about these uncontrolled inputs is usually available in a form of forecasts or historical and/or generated time series representing the most significant uncontrollable inputs. The decision making process cannot take place in an absence of **feedback information** about results of previously applied (selected) controls. This feedback information is based on observations and measurements of the output system variables and state variables.

3.5.4. Challenges Facing the Use of DSS

It is important to mention the challenges that face the development and use of DSS. One of the biggest challenges of the DSS in facilitating access to information by a broad spectrum of stakeholders is associated with the fact, that available information must directly address their concerns and information needs. Therefore it is important to know how the information is obtained from and presented to non-specialists: what information is or should be presented, what form the information has, and how the access to the information is managed. The next challenge is associated with providing non-professionals in technical matters with the possibilities to obtain answers to questions, which are relevant to these groups, especially in a case, when both the questions and responses not necessarily have to be expressed in technical terms. The information presented to non-specialists cant substitute or hide real facts. This information must contain the same value as far as real consequences of considered decision alternatives are concerned, but the form of this information should allow for straightforward recognition of impacts, perils and benefits.

The only possible method to adequately respond to these challenges has been associated with the balanced and targeted usage of technical and technological means combined with such organizational forms of the decision making processes, when also professional in non-technical disciplines and various groups of interest have the right to participate in the evaluation of considered alternatives and their impact.

3.5.5.   Decision Support Systems for Water Resources Management

The water resource issue have so many dimensions and complexities that decision support is necessary for water managers. Kaden et al.(1989) describe a *simple* computer aided water management system, to assist the operators, managers, and planners composed of the following   three components, as shown in figure (3-6):

- Measuring and information systems (Data acquisition, transmission and storage)
- Software system (user software)
- Organizational system (organizational structure, legal and economic regulations)

Haagsma (1995) described a more sophisticated, internet linked Decision Support System.   It should be realized that decisions in regard to water resources are not necessarily taken on the basis of objective information alone, and that political considerations do come into the picture.   A large amount of data has to be transferred into useful information. The aim of the decision support system is to make available the relevant information to both the experts and water authorities, to enable them to have an informed consultation.

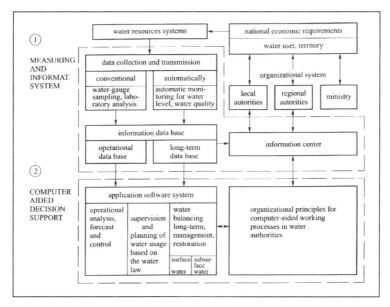

Figure (3-6): Computer –aided water management system. (source: Kaden et al. 1989)

3.5.6. Water Quality Models in DSS Computational Frameworks

In the design and development process of DSS for integrated water resources management and planning, water quality issues are a major component that should be taken into account. Water quality mathematical models are considered essential components of building the computational framework of such DSS's. Modelling, in the decision making process, provides the answers to decision making questions associated with environmental problems. Therefore, several studies were accomplished on either building a standalone DSS for water quality management or on developing the appropriate computational frameworks of DSS's including water quality modelling tools.

Several studies of integrated environmental information and decision support systems are described in the literature. As an example of a developed system; in Lotov et al. (1997) describes the Multi-criteria DSS for River Water Quality Planning which is a computer-based system for supporting the water quality planning in river basins. It was developed on request of the Russian State Institute for Water Management 1993-1994. The integrated mathematical model used in the system consisted of two parts:

1) A collection of wastewater treatment models; models which relate the decrement of pollutants emission to the cost of wastewater treatment in a industry or a service. The models applied the same idea of technological description: a decision variable described the fraction of wastewater, which should be treated by a certain wastewater treatment technology in a given industry or a service placed in a given reach;

2) A pollution transport model which provided an opportunity to calculate concentrations of pollutants at monitoring stations for a given discharge.

Another example of a water resource management DSS including a water quality component, is the system explained by Fedra, (1995) *river basin information and management system* (WaterWare). It is an integrated, model-based information and decision support system for water resources management. Designed for use by water industry and government agencies, it provides the tools to analyze environmental impacts and constraints of water resources management options. Using a set of databases, a geographical information system, simulation models, and expert systems, the software addresses a number of issues such as basic water allocation and use strategies, including the often conflicting requirements of agriculture, industry, domestic use, and recreational and environmental requirements; environmental impacts from resource development projects such as reservoirs; river and reservoir water quality (as a major constraint on the use of water) under numerous land use and point discharge (waste load allocation) scenarios, eg., from industrial and domestic treatment plants but also urban storm water runoff; or groundwater contamination from landfills and its potential impact on drinking water.

Several other examples for DSS's for water resources management and for specific water quality management were developed, such as: BASINS (Better Assessment Science integrating point and non point sources) developed by EPA, 1996 as a multipurpose environmental analysis system designed for use by regional, state, and local agencies in performing watershed and water quality-based studies. DESERT (Decision Support system for Evaluation of River basin strategies), is another sophisticated highly integrated tool for decision support in water quality management on a river basin scale, developed by The International Institute for Applied Systems Analysis (IIASA), Ivanov et al., (1995), Austria. It integrates most of the stages, which can be found in usual decision support procedure: data management, model formulation and implementation, calibration, simulation, optimization and plotting of results. DESERT has a built-in capability for the evaluation of least-cost strategies through the selection of optimal upgrading alternatives of wastewater treatment plants.

The main structure of all these water quality DSS's depends on the structured databases and computational frameworks. These computational frameworks comprise water quality modeling tools of different properties and approaches, which depend on many factors such as the addressed water quality modeling problems, scale of application, parameters of concern and management objectives.

Data is the key issue in developing a reliable DSS that could be used in solving water quality problems and managing water resource. A well structured and designed DSS can still give misinterpretation and wrong information for decision makers if the data used is not with accepted quality. Despite big progress made in recent years with respect to collection and storage of data, including usage of remote sensing technology, availability of reliable, credible and consistent data has been and will remain a problem for next years. Collection and storing of data requires not only technical and technological infrastructure, but also high investment in measurement networks and in processes of data validation and verification. Therefore without significant efforts and investments, which have to be carried out by governments and international agencies, it is hard to believe that any significant process will be achieved in developing reliable DSS for aiding the decision making process of water quality.

3.5.7. Uncertainties in the applied DSS tools

Modeling and decision support tools (e.g., integrated assessment models, optimization algorithms, and multicriteria decision analysis tools) are being used increasingly for comparative analysis and uncertainty assessment of environmental management alternatives. If such tools are to provide effective decision support, the uncertainties associated with all aspects of the decision-making process need to be explicitly considered. However, as models become more complex to better represent integrated environmental, social and economic systems, achieving this goal becomes more difficult.

The new high expectations for the aquatic environment, incorporated into the current wave of regulations, is prompting additional complexity with regard to modeling spatial variability, micro-pollutants and ecological indicators (Somlyody et al. 1998; Thomann 1998). Facilitated by improved computational resources, there is a trend for spatial discretisation to be increased, multi-media models to be developed (e.g. Havnø et al. 1995), and for traditional water quality determinands to be broken down into constituent species (Chapra 1999). As a consequence, the typical number of modelled components has risen exponentially over the past years, and this growth is expected to continue (Thomann 1998). Despite the increasing expectations placed upon water quality models, contemporary deterministic models, when audited, frequently fail to predict the most local and basic biological indicators with a reasonable degree of precision (e.g. Jorgensen et al. 1986). Even when models are claimed to be 'reliable' following audits, a very significant margin of error is allowed (e.g. Hartigan et al. 1983).

Uncertainty analysis for model simulation is of growing importance in the field of water quality management. The importance of this concern is provided by the recent awareness over health risks from improper disposal of toxic wastes as well as continuing emphasis of risk assessment (Radwan et al., 2004). Mechanistic modeling of physical systems is often complicated by the presence of uncertainties, which could be classified into different categories according to the study limitations. Uncertainties in water quality modeling can be classified as Natural, Model, and Data uncertainties. Model input, model structure, model parameters are examples of uncertainties that could be associated with water quality modeling. As presented by Radwan (2002), different approaches for representation of uncertainty are applied; classical set theory, probability theory, fuzzy set theory and rough set theory. The most widely used uncertainty representation approach in transport – transformation modeling is the probabilistic modeling approach.

McIntyre et al., (2003), stated that while additional model complexity might be expected to improve the precision of model results, this has proven to be unfounded in a variety of studies (e.g. Gardner et al. 1980; Van der Perk 1997; also see Young et al. 1996). Furthermore, future driving forces such as climate (Parker 1993) and distributed pollution sources (Shepherd *et al.* 1999) are poorly defined and themselves cannot be modelled with much precision. Clearly, identification of suitable water quality policy must take account of the uncertainties associated with both the validity of the models and the driving forces. However, as increased model complexity hinders the formal evaluation of uncertainty, due to the large number of uncertain model components to be simultaneously analysed, there is a danger that our ability to evaluate uncertainty will decrease. To allow intelligent use of complex simulation models, and to allow informed interpretation and application of model predictions, it is essential that a new generation of tools is developed and disseminated. These should be directed at evaluation of model uncertainty, as well as its minimisation, with respect to the modelling tasks. For results to be justified and interpreted properly, methods used for uncertainty analysis must be theoretically or intuitively well-founded and transparent to the modeller. For methods to be practical for day-to-day use, they should be relatively easy and fast to implement.

Ascough et al., (2008), presented in their study on future research challenges for incorporation of uncertainty in environmental and ecological decision-making, some of the important issues that need to be addressed in relation to the incorporation of uncertainty in environmental decision-making processes and these include: (1) the development of methods for quantifying the uncertainty associated with human input; (2) the development of appropriate risk-based performance criteria that are understood and accepted by a range of disciplines; (3) improvement of fuzzy environmental decision-making through the development of hybrid approaches (e.g., fuzzy-rule-based models combined with probabilistic data-driven techniques); (4) development of methods for explicitly conveying uncertainties in environmental decision-making through the use of Bayesian probability theory; (5) incorporating adaptive management practices into the environmental decision-making process, including model divergence correction; (6) the development of approaches and strategies for increasing the computational efficiency of integrated models, optimization methods, and methods for estimating risk-based performance measures; and (7) the development of integrated frameworks for comprehensively addressing uncertainty as part of the environmental decision-making process.

As model complexity increases in order to better represent environmental and socio-environmental systems, there is a connected need to identify potential sources of uncertainty and to quantify their impact so that appropriate management options can be identified with confidence Ascough *et al.*, (2008). Many studies have focused on the identification and quantification of certain aspects of uncertainty, such as the development of risk-based performance measures (e.g., Hashimoto et al., 1982), and the incorporation of uncertainty into environmental models (e.g., Burges and Lettenmaier, 1975; Chadderton et al., 1982; Eheart and Ng, 2004), optimization methods (e.g., Cieniawski et al., 1995; Vasquez et al., 2000; Ciu and Kuczera 2005), multicriteria methods (e.g., Rios Insua, 1990; Barron and Schmidt, 1988; Hyde et al., 2004), multi-period multicriteria model uncertainty analysis (e.g., Choi and Beven, 2007), decision-support tools (e.g., Pallottino et al., 2005; Reichert and Borsuk, 2005), and adaptive management systems (e.g., Prato, 2005). Only a few research studies have taken an integrated approach that identifies and incorporates all sources of uncertainty into the decision-making process (e.g., Maguire and Boiney, 1994; Reckhow, 1994; Labiosa et al., 2005), and several regional co-operative research efforts are underway to address this issue.

Several research traditions provide concepts, logic and modeling tools with the intent of facilitating better decisions about the environment (Jaeger et al., 2001). Important factors that have an impact on whether and how environmental and ecological problems are addressed are shown in Fig. (3-7). Firstly, environmental problems need to be identified and brought to the attention of managers and decision-makers in the Problem Structuring phase. This can be done through the reporting of routine data, modeling efforts, or input from local stakeholders and/or lobby groups. Once a particular problem is on the agenda, a decision to take action has to be made. This decision will depend on factors such as the perceived importance and magnitude of the problem, as well as financial and possibly political considerations. Following a decision to act, the selection of appropriate assessment criteria and a list of alternative solutions have to be generated.

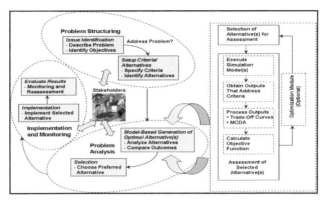

Figure (3-7): Schematic of the environmental decision-making process with an emphasis on model-based decision support tools to generate optimal alternatives

The literature shows that there is neither a commonly shared terminology nor agreement on a generic typology of uncertainties. The recent studies on uncertainties associated with tools used for decision making, showed that many sophisticated approaches to environmental decision-making contain a modeling or some other type of formal decision support component. In this context, modeling for environmental decision support should provide decision makers with an understanding of the meaning of predictive uncertainty in the context of the decisions being made. To a decision-maker, the possible outcomes resulting from a course of action are of main interest, where an "outcome" is defined in terms of the variables of interest to the decision maker.

Predicting outcomes involves the integration of all sources of uncertainty, including uncertainty in model parameter, structure, and output, system variability, decision-making criteria, and linguistic interpretation. These sources of interest can include social and economic endpoints and other variables outside the expertise of ecologists and environmental scientists which often contribute to some of the difficulties associated with transmitting and translating scientific information into policy and decisions. Nevertheless, these variables may be of primary importance for aiding decision-makers in choosing between alternatives (Ascough *et al.*,2008).

4. FRAMEWORK OF THE DEVELOPED WATER QUALITY MANAGEMENT INFORMATION SYSTEM

(FOR DRAINAGE – SHALLOW LAKE CATCHMENT)

Management Information Systems users are not only decision makers. There are a variety of users with different levels of involvement based on authorisation and responsibilities. Decision makers are considered the targeted final users of information produced from such system, for taking policy related actions. Every involved personnel in the management information system is considered a user, whether they are analysts, modellers, managers or decision makers. Each user has a specific role within the application of the system to ensure the flow of information from one level to other until it is used at the highest level for taking decisions related to managing water resources. This chapter focuses on the developed framework for WQMIS for the catchment lake system of irrigated watershed. The developed water quality management information system in this study, with its decision support capabilities, in the form of computational framework, will be used by different levels of users according to the required type and level of information. The system will give a better overview of the water quality situation in the drainage catchments and connected northern lakes, and help to optimise the management of the drainage water quality. Using this proposed user friendly WQMIS with its GIS/RS database, water quality models, will serve decision makers, managers and planners and technical engineers working in the field of water quality management.

4.1. INTRODUCTION TO WATER QUALITY MANAGEMENT INFORMATION SYSTEM WQMIS FOR A CONNECTED CATCHMENT LAKE SYSTEM

Next to quantity, the quality of the water is the most important resources characteristic that needs to be quantified and managed. Water resources cannot be managed without information on both the quality and the quantity being available for a specific purpose. The available quantity is, furthermore, often dependent on the quality requirements of the intended use. Any successfully implemented DSS for water resources management should include a component that focuses on water quality. Research in the field of surface water quality management in general shows that there is considerable complexity in dealing with different processes and types of water bodies. Therefore most ongoing research is oriented to very specific pollution problems, related to certain groups of pollutants and their impacts on specific water bodies. Due to the need to deal with different levels of complexity, when we refer to a surface water quality management information system, we still have to be very specific. In this study the focus is on drainage water from irrigated catchments and its effect on a shallow lake coastal system as explained above. These connected water bodies can only be a part of the water resources existing within a certain watershed. The WQMIS developed in this research is restricted to water quality problems and parameters that lead to eutrophication problems in shallow coastal lakes as a result of poor management of watershed. But still such a developed computational framework is considered a crucial component of a DSS that can be used as a tool for managing the overall water resources within these watersheds.

The main objective of the water quality management information system (WQMIS), which is presented in this thesis, is to serve as a computational framework for decision support in water quality management. This system is developed on a watershed basis, i.e. the approach is to manage a watershed as a whole unit. This means taking into account the various existing water bodies within the watershed, the interrelationships between these water bodies, the water uses and land-use practices, their effects on the water system as a whole, and the different sources of pollution affecting this system. The research focuses on surface water quality management in irrigated agricultural watersheds and specifically the drainage water quality and its impact on connected open water bodies as coastal shallow lakes (the connected-catchment lake systems).

The developed system is composed of three main components; a comprehensive geo-database, a hierarchy of simple and complex hydrodynamic and water quality models; and a set of management scenarios for surface water quality in the catchment-lake system. The system components are described in the coming sections; each component is highlighted, including the geo-database structure, the different data types and categories, and the use of the database to retrieve different levels of information. The conceptual modelling framework has been developed to address each of the objectives concerning water quality management in the connected catchment-lake system. The modelling framework and its significant components are described, including the hydrodynamic models of the catchment and shallow lake, the water quality and eutrophication models. The modelling processes and links between different models are discussed.

4.2. WQMIS COMPONENTS AND STRUCTURE

The proposed water quality management information system for the Edko watershed includes three components as mentioned above. These components are: first, a comprehensive *surface water quality geo-database* including the whole watershed with emphasis on the catchment and lake systems. This database is intended to be the core of the developed information system. It includes all the categories and types of data that are used as either inputs to the modelling components or outputs from the models that could be used for decision support. This geo-database is also designed to be accessible by different levels of users based, for example, on the level of data needed and the associated application. The geo-database incorporates all types of data needed for the development and operation of the water quality models. It includes, hydrographic survey data, GIS layers and base maps, spatially referenced water quality data, satellite images and processed images, spectral water quality measurements,...etc. The second component of the WQMIS is the ***modeling tool***. This tool includes the following modules:

- The first module is a simplified 1D-2D catchment-lake model for the surface network and connected lake system; the model is developed to simulate the hydrodynamic flow of the main drainage system to the lake. This modelling tool is considered the base model and it aims at a better understanding of the hydrodynamic behaviour of the catchment-lake system and the effect of the catchment hydrodynamics on the lake hydrodynamics.

- The second module is a detailed integrated 2D shallow lake hydrodynamic and water quality model, which is considered to be the main model of the lake system. The lake model gives a detailed overview of the lake system hydrodynamics and simulates the spatial and temporal variations of the hydrodynamics and of water quality. The water quality model is divided into two components: a basic indicators water quality model and a eutrophication screening model. The two modules give a complete overview of the water quality condition and eutrophication level within the lake, and in turn they reflect the conditions within the watershed.

The third component in the WQMIS is the ***Remote sensing tool;*** in this developed information system, the role of remote sensing is emphasised for both data acquisition and as a key component upon which the computational framework (modelling) is based. The remote sensing data is applied at all steps of developing the computational framework: from the data acquisition phase for building the geo-database, to the analysis of measured field data and processing of images for developing the algorithms for the extraction of concentration maps of TSM and CHL-a to the use of the developed data in calibrating the water quality models. Therefore remote sensing and associated analysis techniques for water quality are essential tools in this study, and the modelling component is dependent on it. In other words, the remote sensing tool in this system is considered a critical tool for decision support. The three components of the WQMIS are integrated for use as tools in a water quality DSS for the drainage network and the lake system. This system is tested by applying it to different management approaches and scenarios. Figure (4-1) shows the general components of the WQMIS, and the links between the different components.

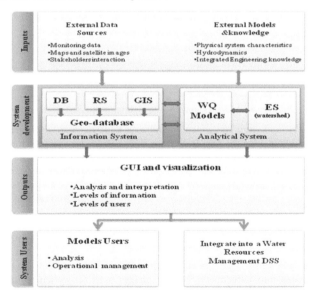

Figure (4-1): The general components of the WQMIS, and the links between the different components.

4.2.1. Surface Water Quality Geo-Database

GIS technology advances over the past several years provide effective and efficient data management for processed hydrographical and hydrodynamic data. The geo-database model lends itself well to database designs and applications associated with the collection, documentation, distribution, and analysis of large amounts of vector, raster, and surface modeling data (i.e., a data warehouse). The geo-database as a database model has two major concepts. First, a geo-database is a physical store of geographic information inside a DBMS. Second, a geo-database has a data model that supports transactional views of the database (versioning) and also supports real-world objects with attributes and behaviour (intelligent features). Behaviour describes how an object can be edited and displayed. The coming section explains in details the structure of the developed geo-database for the WQMIS and its different components.

The water quality geo-database is developed using Arc-GIS software embedded applications (geo-database and visual basic programming tools). The developed system is composed of three main modules to organise and structure the data used in the WQMIS, these modules are: the catchment component, lake component, modelling and analysis outputs. Figure (4-2) gives an overview of the database in Arc-GIS environment.

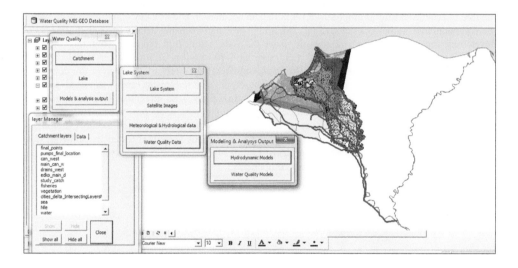

Figure (4-2): Water quality geo-database in Arc-GIS environment.

Since the different modelling components requires a huge numbers of preparation input files and a large number of output files the geo-database is designed to store and retrieve all types of data files needed for the development and operation of water quality models. It includes, hydrographic survey data, GIS layers of the system under investigation, spatially referenced water quality data, satellite images and processed images, spectral water quality measurements,...etc. Outputs from different models are stored within the developed system to be retrieved at any time for analysis or interpretation.

The different parts of the database aims at structuring the different needed datasets for each modelling component, therefore the catchment part includes the main GIS layers that are used in the 1D-2D hydrodynamic model and the associated datasets needed for the model inputs. For the lake part the datasets are divided into GIS layers, satellite images used for system delineation and for models calibration at different stages, and the meteorological and hydrological datasets needed for the detailed hydrodynamic model, and the water quality historical data and measured data linked to geographic locations of monitoring stations and field work selected measuring sites. The modelling and analysis part focuses more on the different models and the detailed parameters and inputs to the models (1D-2D), detailed 2D hydrodynamic and water quality and eutrophication model.

4.2.2. Remote Sensing Of Water Quality Parameters

The use of remote sensing techniques and tools in this research work is an important key component in the framework of the management information system. Due to the very limited temporal ranges of monitored water quality data and the neglected spatial and temporal monitoring of important water quality parameters (such as TSM and CHL-a) that are essential for eutrophication modelling, environmental management in the catchment area specially the coastal lake (and similar areas within the Nile delta region) this catchment can be referred to as "data scarce environment". In such environments where there is a serious lack of essential data for management purposes, the remote sensing can be used as a powerful tool to compensate for (to a reasonable extent) missing needed datasets. Time series water quality data sets can be derived from remote sensing satellite images. Also spatially distributed water quality data sets can be a useful product from remote sensing analysis of satellite images. In addition to these important uses of remote sensing, the extracted water quality parameters can be used for forecasting and prediction applications.

4.2.3. Mathematical Modelling of Hydrodynamics and Water Quality

The major components in the framework of the WQMIS are the mathematical models including the 1D and 2D hydrodynamic models of catchment lake system, the detailed 2D hydrodynamic model of the shallow lake system, the basic parameters water quality model and the eutrophication screening Lake Model. This hierarchy of models is considered the backbone of the watershed management system. Figure (4-3) shows the mathematical models components and the links between them. Chapter 7 focuses on all the modelling components and the links between these models.

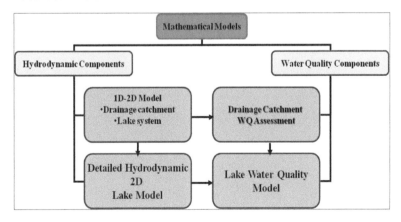

Figure (4-3): Mathematical models components and the links between them

4.2.4. Decision Making Tools to Explore Scenarios and Reflect Decisions for WQ Management

In general a DSS for Water Resources Management consists of three different layers:

- An information system, including databases, GIS, etc.
- A computational framework, with a simulation model (s), (water quantity, water quality, lake, waste load, etc.)
- An analysis layer, where optional Water Resources Management strategies under various scenarios are defined and analysed on their impacts on water issues, overall environment, and socio-economics.

This research study focuses on the development of the computational framework of the decision support system for surface water quality management which includes the mathematical models, remote sensing tools and the geo-database including all datasets used in modelling. The water quality and eutrophication physically based models are the main developed decision support tools in this research as a part of the computational framework developed for the catchment-lake system. These two models are considered the sub components used for decision support. Different management scenarios for reduction of pollutants entering the lake can be investigated using these models. Since agricultural drainage is the main source of pollution to the lake, introducing excess amounts of nutrients in to the lake system, proposed scenarios of nutrient rates reduction will be investigated for managing the lake system as shown in chapter 9.

5. DESCRIPTION OF THE STUDY AREA AND DATA INVENTORY

The study area selected for application of the proposed water quality management information system id Edko drainage catchment and lake in the western Nile Delta region. The main reason for selecting this watershed is that it is a typical irrigated watershed with physical complicated nature, and it involves different water quality problems in the catchment and lake area. This chapter gives a detailed overview of the study area and the different categories of data used in this research. It ends with a preliminary analysis of the collected data and primary water quality assessment of the study area.

5.1. INTRODUCTION

The Nile Delta and the Nile river valley of Egypt is one of the oldest agricultural areas in the world, having been under continuous cultivation for at least 5000 years. The arid climate of Egypt, charaterised by high evaporation rates (1500-2400 mm/year) and little rainfall (5-200 mm/year) leaves the river as the main fresh water supply. Under these arid conditions, no natural soil development can take place, figure (5-1) shows the Nile Delta region and Nile valley. The river Nile, therefore, is the life of the country providing:

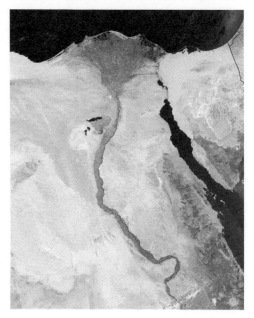

- Fresh water supply for agriculture, industry and domestic use;
- Hydro-electric power generation;
- Navigation

Available information shows that the river Nile, its branches, canals and the drains are suffering from an alarming increase in pollution through wastewater. Total Dissolved Solids (TDS) range from 130 mg/l in Lake Nasser to 200-250 mg/l at the Delta barrages. The pH increases from 7.7 at High Aswan Dam to 8.5 in the Nile Delta. The BOD as a result of human activities mainly shows a random distribution but only occasionally exceeds the standard of 6 mg/l (especially in the downstream sections). The random distribution is the result of point discharges and self-purification of the river.

Figure (5-1): Nile river valley and delta

As a result the dissolved oxygen only drops in exceptional cases below the threshold of 5 mg/l. Nitrate and ammonium show hardly any exceedence of the current standards, except for ammonium at one certain locations. The spatial distribution of faecal coliform varies strongly. The standard is significantly exceeded during the summer months at a few locations in Upper as well as Lower Egypt. It is most important that the Nile maintains its self-purification capacity. However, enormous loads of pollutant matter are released to the system. To what extent the Nile sediment is contaminated with accumulated constituents is not known.

5.2. THE PRESENT STATUS OF THE NILE DELTA AS AN IRRIGATED AGRICULTURAL WATERSHED

The Nile Delta extends between Latitudes $30°10'$ N and $31°30'$ N about 23 Km. To the north of Cairo, where the Nile Valley opens out to form a triangular area of about 49600 km^2. This area is considered one of the most fertile and intensively cultivated regions of the world. Many centuries ago seven branches of the Nile used to flow through the delta, but during time and different geological eras, many geological changes and formations have taken place and the Nile river has been diverted into two main branches, the Rosseta branch on the west side of the Delta, and the Damietta branch on the east side. The Delta Barrages about 26 Km. North of Cairo, control the water supply to these two branches. The Delta has a slope of about 9 cm./km. To the north west, it is bordered by higher deserts from east and west. On the northern estuarine boundaries a series of shallow lakes occupy a great part of the northern section of the Delta. These lakes receive a considerable amount of drainage water from the Delta, and are separated from the sea by narrow strips of land and have outlets to the sea, (Amer et al., 1989). Figure (5-2) shows the map of the Nile Delta.

The Damietta Branch begins at the Delta Barrage and ends 220 km downstream at the Faraskour dam near Damietta. The Rosetta Branch also starts at the Delta Barrage and flows to the Mediterranean at Rosetta. Deterioration in water quality of the two branches does occur in a northward direction due to the disposal of municipal and industrial effluents and agricultural drainage as well as increasing pollutant concentration due to decreasing flow.

The Rosetta Branch receives high oil and grease concentrations, nutrients, organic loads, and solids. This is a result of the discharge of a part of the wastewater of Greater Cairo through the Muheet and Rahawy drains as well as some discharges of pesticides and toxic chemicals from other sources. Also, salts, suspended matter and herbicide residue reach the river from agricultural drains.

The Damietta Branch receives nutrients, organic loads, grease and oils as a result of discharges from the Talkha fertiliser industry and drainage of herbicides and pesticides from agricultural drains especially near the Faraskour dam. total dissolved solids (TDS) increases in the branches up to approximately 500 mg/l.

Irrigation canals have hardly been referred to with respect to water quality monitoring. The canals have a water quality similar to that at the point of diversion from the Nile. The flow in the canals varies with irrigation demands. Most of these canals are major sources for drinking water treatment plants. However, many canals are suffering from the following discharges:

- Industrial and domestic wastes (liquid and solid) from canal banks, as is the case in the Mahmoudia and Ismailia canals.
- Residuals from fertilisers, molluscicides (snail killer, for instance for the control of Bilharziasis) and herbicides which find their way to the irrigation water system.
- A mix of agricultural, domestic and industrial wastewater at locations where reuse pumping stations add drainage water to the canals.

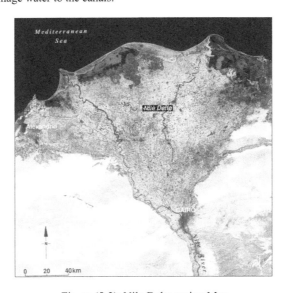

Figure (5-2): Nile Delta region Map

5.3. THE NEED FOR DRAINAGE WATER REUSE IN NILE DELTA

Due to water scarcity, reuse of drainage water is becoming an increasingly important water source in Egypt. However, large quantities of water in the drainage network cannot be used as they contain high contaminant loads. In the Delta region the amount of agricultural drainage water reuse was estimated in 1995/96 to be around 4.27 BCM in addition to about 0.3 BCM lifted to Rossetta branch from west delta drains. This constitutes the official reuse carried out by pumping stations of the Ministry of Water Resources and Irrigation (MWRI). Additional unofficial reuse done by farmers themselves, when they are short of canal water, has been estimated to be around 2.8 BCM.

The remaining drainage water is discharged to the sea and the northern lakes via drainage pump stations. The total amount of drainage water that was pumped to the sea during the year 1995/96 has been estimated to be 12.41 BCM. Reuse of agricultural drainage water in the Delta is limited by the salt concentration of the drainage water. Moving from upstream to downstream, the level of salinity increases but in most of the valley and in the southern part of the Delta region, the salinity remains below the critical level of 1,000 ppm making it possible for reuse. However, in the northern part of Delta region, large quantities of salt seep through groundwater to the drainage water due to the sea water intrusion. The amount of seawater that seeps into the drains is estimated to be about 2.0 BCM/year. This water is pumped back to the sea and northern lakes to maintain the salt balance of the system.

Sectors and Sources contributing to Water Pollution

Degradation of water quality is a major issue in Egypt. The severity of present water quality problems in Egypt varies among different water bodies depending on: flow, use pattern, population density, extent of industrialization, availability of sanitation systems and the social and economic conditions existing in the area of the water source. Discharge of untreated or partially treated industrial and domestic wastewater, leaching of pesticides and residues of fertilizers; and navigation are often factors that affect the quality of water in different water bodies.

1. Industrial Wastewater

The industrial Wastewater The industrial sector is an important user of natural resources and a contributor to pollution of water and soil. There are estimated to be some 24,000 industrial enterprises in Egypt, about 700 of which are major industrial facilities. The spatial distribution of industry in Egypt is influenced by the size of the employment pool, availability of services, access to transportation networks, and proximity to principle markets. The manufacturing facilities are therefore often located within the boundaries of major cities and in areas with readily available utilities and supporting services. In general, the majority of heavy industry is concentrated in Greater Cairo and Alexandria. Industrial demand for water in the year 2000 has been reported to be 3.6 BCM/year. By the year 2017, the industrial demand for water is expected to reach 5.5 BCM/year. Consequently, a corresponding increase in the volume of industrial wastewater is expected, (NWRP, 2002).

2. Municipal Wastewater

Municipal Wastewater Based on the population studies and rates of water consumption, the total wastewater flows generated by all governorates, assuming full coverage by wastewater facilities is estimated to be 3.5 BCM/year. Approximately 1.6 BCM/year receives treatment. By the year 2017, an additional capacity of treatment plants equivalent to 1.7 BCM is targeted (National Water Resources plan, 2002). Although the capacity increase is significant, it will not be sufficient to cope with the future increase in wastewater production from municipal sources and therefore, the untreated loads that will reach water bodies are not expected to decline in the coming years. The constituents of concern in domestic and municipal wastewater are: pathogens, parasites, nutrients, oxygen demanding compounds and suspended solids. Therefore, instances of high levels of toxic substances in wastewater have been reported. As these toxic substances (heavy metals & organic micro-pollutants) are mainly attached to suspended material, most of it accumulates in the sludge. Improper sludge disposal and/or reuse may lead to contamination of surface and ground water. In general, the bulk of treated and untreated domestic wastewater is discharged into agricultural drains. Total coliform bacteria reach 106 MPN/100 ml as recorded in some drains of Eastern Delta. It is important to mention that all drains of Upper Egypt flow back into the Nile. Moreover, it has become a national policy to maximize the reuse of drainage water by mixing it with canal water. Many irrigation canals may be contaminated with pollutants from domestic sources as a result.

3. Agricultural drainage water

These non-point sources are, however, collected and concentrated in agricultural drains and become point sources of pollution for the River Nile, the Northern Lakes, and irrigation canals in case of mixing water for reuse. Moreover, these non-point sources of pollution may also influence the groundwater quality. Major pollutants in agricultural drains are salts; nutrients (phosphorus & nitrogen); pesticide residues (from irrigated fields), pathogens (from domestic wastewater), and toxic organic and inorganic pollutions (from domestic and industrial sources).

5.4. THE PRESENT STATUS OF THE DRAINAGE NETWORK AND THE NORTHERN LAKES

5.4.1. Drainage Network

The open drain system receives the excess of irrigation water that flows through the soil or via a constructed underground drainage system. The quality of drainage water is affected by the chemical composition of the soils, toxic substances used for pest or herb control and domestic effluents from the banks. Most of the drainage system of Upper Egypt discharges the wastewater in the river Nile, while most of the drains in the Delta ultimately discharge into the Northern Lakes mentioned earlier.

The agricultural drainage of the southern part of Egypt returns directly to the Nile River where it is mixed with the Nile fresh water and reused for different purposes downstream. The total amount of such indirect reuse is estimated to be about 4.07 BCM/year in 1995/96. This drainage flow comes from three sources; tail end discharges and seepage losses from canals; surface runoff from irrigated fields; and deep percolation from irrigated fields (partially required for salt leaching). The first two sources of drainage water are of relatively good quality water. The deep percolation component is more salty and even highly saline, especially in the northern part of Delta, due to seawater intrusion and upward seepage of groundwater to drains.

Apart from being the largest consumer of water, agriculture is also a major water polluter. Saline irrigation return-flows or drainage containing agrochemical residues are serious contaminants for downstream water users (World Bank, 2001). The drainage system particularly receives the heaviest pollution loads. The major sources of water pollution are municipal and rural domestic sewage, industrial wastewater and agricultural chemicals [salts, nutrients and pesticides]. Overall the agricultural drains receive the bulk of the treated and untreated domestic pollution load. But all drains in Upper Egypt flow back to the Nile, and additionally it has become the national policy to maximize the reuse of drainage water by uplifting it to canals. As a result and as mentioned earlier many canals are now also contaminated with domestic pollutants. This is the present situation of the surface network in the Nile Delta, where dependence on the Nile system makes management of its quality as important as management of its quantity.

Focusing on the situation in the Nile Delta, which is the downstream major consumer of the Nile water, in order to manage its surface water quality, there is a need to deal with a very complicated system. This system is composed of irrigation and drainage networks, which are directly connected to a Northern Estuarine System, agricultural lands, communities, urban cities, different types of land use and various sources of pollution affecting the system, and point and non-point sources. The drains in Delta have extremely variable quality. On the western edge of the Delta, drainage water from newly reclaimed lands receives increased salt concentrations. Some of the Delta drains receive municipal and industrial wastes in addition to agricultural drainage water. A large part of the drainage water is disposed into the Northern lakes and consequently into the sea.

5.4.2. Northern Lakes

Most of the drains in the Nile Delta discharges into lakes at the seacoast, and they, in turn, discharge to the Mediterranean Sea. From east to west there are six terminal lakes, four of which are fed mainly by drainage water with low salinity, these are: Lake Manzala, Burullus, Edko and Mariut. These all have free connection to the open sea except Lake Mariut, which is permanently cut off from the sea. All the lakes are shallow with an average salinity that ranges from 3 to 20 ppt (parts per thousands). The remaining two lakes, Port Foad and Bardawil, are supplied only by seawater.

Part of the drainage water in Egypt is being reused for irrigation purposes by mixing with fresh irrigation water, or diverted and used in land based aquaculture or fish farms. The principal portion of the drainage water is directed to the northern lakes or to the sea; this estimated amount is 12.3 billion cubic meter per year. The drainage water reuse, as a solution for irrigation water deficit, should not decrease the drainage water flowing to the north estuarine region so as not to affect the salt balance, the fisheries and the water quality. These impacts depend on the quantity and quality of drainage water that is discharged to the northern lakes or to the Mediterranean Sea. The decrease in the drainage flows in the North Delta drains due to the implementation of more reuse projects and irrigation improvements will cause further deterioration of water quality, (Imam and Ibrahim, 1996a).

Since salt loads removed from the agricultural land will be nearly the same and the reuse of drainage water in another irrigation application will lead to further salt build up in the drainage water. Similarly, pollution loads discharge from municipal and industrial facilities will be mixed with less drainage water, thereby, leading to higher concentration of pollutants.

The northern lakes at the present time are highly polluted and considered as a sink for many pollutants such as heavy metals, pesticides, fertilizers, nutrients, and pathogens, which enter the drains through the effluents of industrial and domestic waste water in addition to agricultural drainage water. These lakes provide a natural environmental treatment to a considerable portion of the current contaminant loads through sedimentation, biodegradation, plant uptake and other mechanisms, (Imam and Ibrahim, 1996a). Therefore the lakes could be considered as a protection barrier from contamination for the beaches and coastal zones from contamination. Consequently, any further reduction in the drainage water quantity or any further degradation in the quality would increase concentrations of pollutants leading to a stressful environment. The overall ability of these lakes to reduce pollutants will be damaged and may lead to an increase in the number of pollutants and contamination to the surrounding coastal zone.

The northern lakes provide a rich environment for estuarine and marine fish. They are the most important areas for fish production in Egypt. Now, they contribute over 50 % to fish sold in Egypt. Fish ponds are not allowed to use canal water due to water shortage, but they can use drainage water. The drainage water has the advantage of containing fertilizers which stimulates the primary production of the fish ponds and consequently increases their production. On the other hand, the drainage water is polluted with different kinds of pollutants, which are considered a risk to the fish and consumers. The quality of water supply for the fish and aquaculture in the lakes is a very important variable, as it mainly governs the biological productivity and the species, which can be reared. The productivity of the water depends on the nutrient inputs in the water and the ability of the ecosystem to utilize these loads. The other main variable is salinity. There is much evidence to show that differences in salinity tolerance lead to the segregation of fish species.

The five lakes bordering the northern coast of Egypt, together represent 25% of the remaining wetland habitat in the Mediterranean basin. Residents of these lakes traditionally exploited a wide variety of resources. Today these lakes face a number of threats to their existence, including large-scale reclamation and water pollution besides global warming and sea water level rise. Agricultural authorities, engineers, fishery managers, and conservationists in Egypt and abroad debate about how best to manage and develop the lake region's resources, but few of these groups understand or communicate with one another, or with residents of the lake communities (Parmenter, 1991). To study such an interacting system of drainage networks and lakes suffering from water quality deterioration, the integrated watershed approach should be taken into consideration. This means that the Delta region should be divided into several sub-watersheds contributing to the three main components of the Delta region (Western-Middle–Eastern). These watersheds could be defined according to the outlet points which are the northern lakes, which in turn are directly affected by upstream conditions. This means that each sub-watershed should be managed as a complete system taking into consideration the northern lakes affected by the drainage water discharged from the upstream catchments.

5.4.3.   Nile Delta Northern lakes Ecology

Vegetation

The northern lakes of the /Nile Delta are considered a rich aquatic environment. The Nile Delta was once known for large papyrus (*Cyperus papyrus*) swamps, but papyrus is now almost absent from the delta. Vegetation consists of *Phragmites australis, Typha capensis*, and *Juncus maritimus*, with some small sedges. The Manzala, Burulus and Edko coastal lagoons supports beds of *Ceratophyllum demersum, Potamogeton crispus*, and *P. pectinatus* around the southern shore as well as dense phytoplankton. Other typical species found here are *Najas pectinata, Eichhornia crassipes*, and *Cyperus* and *Juncus* spp. that grow along lake shores. The salt tolerant *Halocnemum* spp. and *Nitraria retusa* grow in marshes along the Mediterranean coast. Farther south along the Nile river, dense swamp vegetation grows unchecked without the seasonal fluctuations of the river, held back by the Aswan Dam. Figure (5-3a,b,c,d) shows the dominant types of vegetation in coastal lakes and Delta wetlands.

Figure (5-3a): Ceratophyllum demersum   Figure (5-3b): Phragmites Australis

Figure (5-3c): Potamogeton pectinatus   Figure (5-3d): Eichhornia Crassipes

Fisheries

The Nile delta in Egypt is typified by a very flat topography. This has permitted the development of shallow fish ponds bounded by earth dykes, both on the strips of land between the Mediterranean sea and the coastal lakes (coastal howash is the common name), and within the lakes themselves (lake shore and lake water howash). Tang (1977) reports that coastal howash typically have a salinity between 10 and 25 ‰ and that these waters are affected by tidal action. Lake shore howash normally have a salinity of less than 5‰ and their water level is controlled by the discharge of irrigation canals or by pumping (personal observation). Lake howash have a salinity varying with location, and may have a water depth of up to 2 m. These howash are generally smaller than the others.

Howash management is very simple. Between May and August, when the water level of the coastal lakes is high due to increase of water discharge from the Nile irrigation system, the howash are opened. Seeding takes place naturally, with tilapia, grey mullet, catfish, perch, eel, etc. The growing period is short, as all irrigation canals are closed in January and February, and harvesting is carried out from November to January. Commonly, poultry manure is used for fertilization, resulting in average yields of 1475 kg/ha in the Lake Manzala region (Tang 1977). Because of the short growing period, however, only 360 kg of this yield is suitable for human consumption, smaller fishes being used for animal feeds. The value of the crop is, therefore, greatly reduced. With the use of supplemental feeds, production of 3.4 t/ha are reported.

5.5. SELECTED PILOT WATERSHED (EDKO CATCHMENT- SHALLOW LAKE SYSTEM)

5.5.1. Description of the Study Area

Edko drainage catchment and lake are selected as the components of the study watershed, to develop the proposed integrated modelling system that could be applied to irrigated watersheds. This watershed is a part of the Western Delta region. It is considered one of the main drainage catchments in the western delta. It serves the North-Eastern part of the Western Delta. This watershed comprises a total area of about 2048 km^2 including the area of the catchment and the northern lake Edko. The watershed is located in the eastern part of the Western Delta region.

Figure (5-4) shows the Western Delta Region and the study area. The eastern edge of the area is Rosetta Branch of the Nile. The southern edge of the area is the Noubaria Canal, and from the west the area is bounded by Khairi drain. The north boundary of the watershed is Lake Edko.

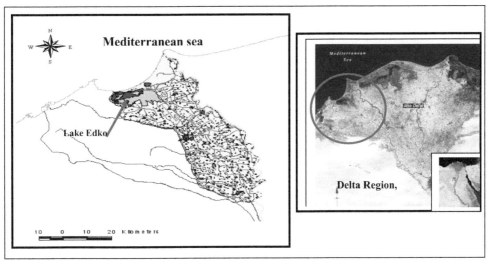

Figure (5-4): Layout of study area

5.5.2. Watershed Characteristics and Main Components

The study watershed is an irrigated watershed, which comprises a drainage catchment connected to a shallow coastal lake as mentioned above. To define the watershed and its detailed characteristics for water quality management a number of items and properties should be taken into consideration. These properties will be the main factors affecting the analysis and modelling processes of the water quality parameters in different water bodies within the watershed. The following items represents the main characteristics of the study watershed;

Edko watershed is divided physically into two main components: the drainage catchment and the shallow lake system. The drainage catchment is composed of Edko main drain and Barseek drain collecting drainage water from the whole catchment into Lake Edko. The lake shallow system is divided into three main components; the aquaculture belt around the lake vegetation, the vegetation belt, and the free surface water body of the lake connected to the see outlet channel. Figure (5-5) shows the watershed components.

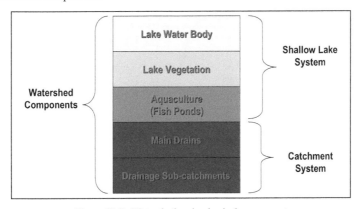

Figure (5-5): Watershed main physical components

91

The Drainage Catchment System

The drainage catchment consists of main drain Edko, which collects drainage water from eight sub-drainage networks from the southern and eastern sides of the catchment and Barseek drain which collects drainage water from the western part of the catchment. The area of the whole catchment is about 2031 km². Average length of catchment is 75 km and average width is 25 km. The soil properties in the drainage catchment are mainly the same soil type as in all the Nile Delta, heavy thick clay, with higher proportions of silt near the northern edges. The main drain Edko has a length of 48.8 Km starting from Shubrakeet pumping station (km 48.8) and ending at the mouth of Maadia Bougaz (Km 0.0) at the inlet to the sea. The designed bed width is about 50.0 m and designed bed level is about –5.0 m.

Six pumping stations feed the drain along its course with drainage water which then flows by gravity to Lake Edko. In addition, there is a pumping station that takes water off the drain for irrigation called Edko irrigation pumping station which is located near the upstream part of Edko drain. On the southern boundary of Lake Edko, there is Barseek pumping station that discharges its water directly to the lake from Barseek sub-catchment. If the water level in Edko drain is higher than this permissible water level the efficiency of the pumping station is greatly reduced to the point of forcing the pumping station to shut down.

The water quality in the drain is monitored on monthly basis by the Drainage Research Institute (DRI). Edko watershed is considered a multiple zone watershed, i.e. comprised of sub-watsheds discharging water into the main drain. In the western delta there are a total number of 35 monitoring locations out of a total number of 113 monitoring points in the whole Nile delta. These monitoring points are divided into: pumping stations (22), open drain locations (11), and outfall locations (2). In the selected study area 13 monitoring locations are present, 12 of which monitor the Edko drain and one station monitors the Barseek drain. Figure (5-6) shows the layout of the catchment .

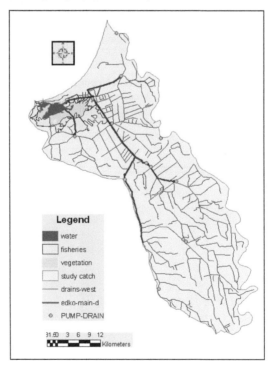

Figure (5-6): Layout of the drainage catchment

The History of Lake Edko

Before going into the present status of the lake and its ecosystem, this section gives some historical information on the lake area and its characteristics. The lake is a typical delta edge coastal lagoon lake with a history of steady contraction from more than 200 km^2 before the 1940s to some 150 km^2 in the 1950s to the present 70 km^2 (Fisheries Department estimates). It is estimated that a reduction of some 30 percent of the lake area has occurred in the last 20 years by the development of drainage and irrigation schemes in the eastern portion. In addition to the losses due to agricultural use, the changes brought about by increasingly intensive agriculture, by lowered quantities of available water and by greater economy in water use for irrigation, have lead to an acceleration in the growth of reeds (mostly Phragmites) and more recently the development of floating masses of water hyacinth (Eichhornia). The lake also supports considerable stands of sub-aquatic plants (Potamogeton), with these occupying a possible 50 percent of the apparently open water surface.

Typically this is a shallow, eutrophic lake at an advanced and accelerating stage of senescence, or evolution to dry land. The average total of 100 km^2 can be assumed to be the productive area for the assessment of possible available fishery resources. The lake occupies a very shallow basin with little variation in a depth between 0.5 and 1.2 m. Encroachment of reeds and deposition of silt reduce the margins to shallow marshes, which allow reclamation and encroachment by agriculture, particularly in the south. To the north the lake boundary is fixed by the coast road and railway line which follow the line of the coastal sand dunes. There is some encroachment of the lake margins on this northern edge. In the east the lake edge is defined by the Edko Drain carrying irrigation drainage raised from the neighbouring agriculture areas. Apart from the Edko Drain, the Barseek Drain enters from the middle of the southern shore. The lake drains to the Mediterranean Sea through a single narrow channel at the sea fishing port of Maadia. It is understood that occasionally this drainage reverses with the inflow of sea water into the lake. There is evidence of this from the response of the vegetation in the vicinity of Maadia and the area of the lake to the west. The drainage of agricultural, irrigated land into the lake provides a nutrient-rich input with a salinity of about 2–3 ppt over much of the lake surface.

The water is not homogenous in appearance, varying from clear to at least the maximum 1.0 m depth, to very cloudy with sediment/plankton and in some areas foul anoxic water smelling of hydrogen sulphide. Such patchiness in the water mass would be expected due to the low flows and the localized effect of water passing through, or being blocked by, stands of submerged plants or where reversal of current flow, through tidal or wind effects, causes water to be flushed out from below consolidated stands of water hyacinth. There is apparently no significant input of organic or industrial pollutants to the lake except for untreated domestic waste water dumped into the drainage network, apart from the enhanced nutrient status of the discharged drainage waters and the residual load of pesticides they carry. There is however a considerable input of herbicides (currently Ametryn) in the October of each year for the last 10 years as a part of a programme to eradicate or reduce the water hyacinth. It is not known what effect these chemicals and their residuals will have on the fish or the edible quality of the fish flesh. It appears that there is no monitoring programme for pesticide residues in the waters of the area or edible products derived from them.

The lake is exposed to the prevailing winds with a reach over the sea or the flat agricultural lands of the delta. With its shallow depth and relatively large area, it is unlikely that there will be any significant stagnation or stratification. The exception to this will be those areas which are blocked by floating and emergent weeds. These are localized and any deterioration in water quality would provide little risk to the fish populations. Discrete water bodies are observable within the lake by colouring and clarity, but their overall pattern could not be ascertained. The major significant heterogeneity is the area to the west of the connection with the sea at Maadia. The lack of aquatic vegetation over a considerable area suggests that this part of the water body is subject to considerable fluctuations in salinity, with presumably the salinity changing with the reversal of flow through the lake-sea junction.

Some historical maps of the lake area were found in literature, Figure (5-7) includes three of these maps at different years. The maps reflect the different conditions of the lake ecological system within the past century.

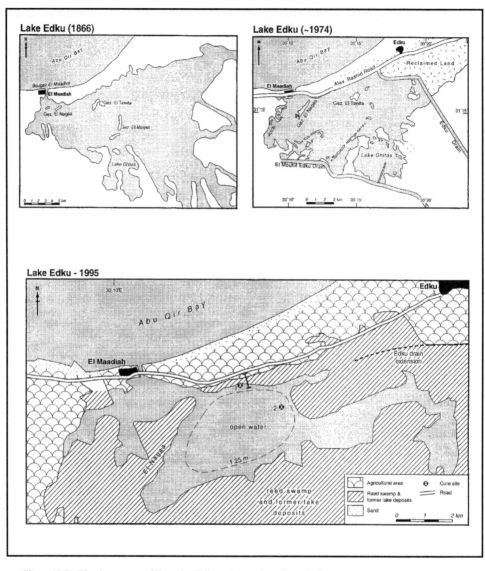

Figure (5-7): The three maps of Egyptian Edko Lake on the Nile Delta indicate lake status on three periods, 1866, early 1970s (after Samman, 1974) and in 1995.

The maps illustrate the scales and diminishing area of the lake water body within time. They also shows the transition of the reclaimed areas into commercial fish ponds that surrounds the vegetation and the water body. In the coming section of this chapter describing the data analysis and assessment more information on the lake status is shown through satellite images and change detection comparison between different satellite images.

The Shallow Lake System current status

The Edko lake system could be defined as a *wetland system of a total area about 124 km²* estimated from recent satellite images (ASTER 2004). This wetland system is composed of a shallow lake water body surrounded by a water vegetation belt and a belt of aquaculture fish ponds. Satellite images taken in 1981 indicated a loss of more than 20% of the water body through land reclamation since 1923.

1. Water body

Lake Edko is a shallow natural brackish water body situated at latitude 30° 15' North and longitude of 31° 15' East of Alexandria city, Egypt, and it extends about 19 km south of Abu-Kir Bay from the east to west. The Lake has an average water depth between 0.8 m to 1.2 m and bed level of –0.5 m. It is connected to the Mediterranean sea with a channel (Maadia Bougaz) of about 50.0 m width with bed levels as deep as –6.32 m. From recent satellite images (ASTER, 2004), the calculated water body area was about 19 km², with average length of is 4 km and average width is 6 km. The deepest parts of the lake are found at the centre of the lake where it reaches 1.4 m. the shallower areas are restricted at the lake, margins, around the vegetation islands and near the outlet channel to the sea. The bottom sediments of the lake are composed mainly of clay with small proportion of sand–sized sediments mainly of calcareous shell fragments and plant debris.

2. Lake vegetation

Lake Edko is rich in vegetation of hydrophytes and amphibious, and contains several numbers of chrysophyta, cyanophyta, and CHL-aorophyta (El Sarraf, 1976). The approximate total area of the vegetation belt surrounding the water body is 38.66 km². Inside the water body there exist a huge number of constructed floating vegetation mats of water hyacinth and reeds collected by fishermen for fixing the fishing nets in addition to a huge area of fishing nets blocking free navigation inside the lake. According to a study conducted by FAO (1985) on marcrophytes, weed control and use of aquatic plants in Lake Edko, 23 aquatic macrophytes are present in Lake Edko. This is based on plants collected during the investigation. The most important species are the floating water hyacinths: Eichhornia crassipes, the emergent reed Phragmites australis, and the submersed Potamogeton pectinatus and Ceratophyllum demersum. Figure (5-8 a, b) shows the dominant types of vegetation in the lake.

Figure (5-8a): Reeds , the dominant macrophyte in lake Edko (2005)

In areas where the water is less than 1 m deep, the reed <u>Phragmites</u> <u>australis</u> is, and was as far as human memory goes, the dominant macrophyte. Since 1960s, water hyacinth (<u>Eichhornia</u> <u>crassipes)</u> has covered large areas, particularly away from the tidal influence. Submersed plants, principally <u>Ceratophyllum</u> <u>demersum</u> and <u>Potamogeton</u> <u>pectinatus,</u> densely cover large areas of the less saline sections of the lake i.e. the eastern areas of the lake upstream of the coastal road which crosses the lake at it narrowest cross section.

Figure (5-8b): Water hyacinth and submerged vegetation in Lake Edko (2005)

3. Aquaculture and Fisheries

The third component of the lake system is the intensive aquaculture around the lake represented by a huge area of fish ponds surrounding the water body, which is estimated to comprise an area of 66.38 km². This area of fish ponds is a closed system which receives water from the main drains and the lake and discharges water back to the drainage system and the lake. Most of the fish ponds are of small area that ranges between 1-1.5 feddans and use small pumps for recycling water in and out of the system. One large area of fish farms is in the southern part of the lake, bounded by Barseek drain and of area about 3000 feddans. A big pumping station serves this area for water recycling. The main fish species existing in the lake are; Tilapia, Mullet and Carps. Fishing in Lake Edku is an unhealthy activity. About seventy-five percent of the 5000 full-time and part-time fishermen in the lake suffer from schistosomiasis and other parasites. The problem is exacerbated by the dense growths of emergent and submersed plants that prevent water movement and provide a habitat for parasite hosts such as snails.

The following part gives more detailed properties of the fish farm of Barseek being the biggest one constructed in the area and which has direct effect on the lake water quality.

Basrseek Fish Farm

Construction of the Barseek Fish Farm began in 1979. The farm was constructed on a 800 ha site in the southeast sector of Lake Edko, to designs elaborated by the Irrigation Department. Before construction began the site was a mixture of areas of <u>Phragmites</u> reed swamp in the north and west and a drier salt marsh scrub in the south and east of the site. The farm consists of approximately 685 ha of water surface, with 625 ha consisting of 104 production ponds of about 6 ha each with a water depth between 1 –1.5 m. In the southwest corner of the site are a further series of 50 smaller (+ 1 ha) ponds for breeding and raising brood stock. The farm does not include a hatchery, but relies on Government hatcheries at some 30 km distance for fish for stocking.

The water supply is obtained from a land drainage canal and pumped via a pumping house supplied with 5×75 hp electric pumps with a total capacity of 5 000 m³/h. A second pumping house of the same design and capacity drains the whole pond area to the Barseek Drain and eventually discharges to Lake Edko. All the ponds are independently filled and drained via large canals. The inflow of water through the sluices into many of the larger ponds is restricted by the small diameter of the inflow pipes (+ 0.2 m). The farm possesses labour lines, administrative, maintenance and storage buildings, all situated on a site in the southwest corner. The ponds are rectangular, 1.0–1.5 m deep, each provided with independent inflow and drainage sluices. A fish sump is dug around the edge of the production ponds. Most bunds are accessible by vehicle.

The stocking programme in Barseek fish farm involves stocking with grass carp on first filling the pond in October/November or December, after the post-harvest preparation. Mullet fry (mixture of M. cephalus and M. capito) are added in March when these become available from the coastal waters. In May common carp are stocked. Tilapia is mostly stocked by uncontrolled inward migration from the canals. Harvesting takes place in October to February. The small size of the mullet and carp implies a low survival rate and presumably the ponds are stocked well below their carrying capacity towards the end of the growing season. The management of both the capture fisheries of Lake Edku and the Barseek Fish Farm are under the jurisdiction of the Department of Fisheries of the Governorate of Beheira. The discharge of the Barseek fish farm into Lake Edko through Barseek drain is an obvious water quality problem affecting the lake. Figure (5-9a) shows the fish ponds around the lake area and figure (5-9b) shows the pumping stations of Barseek fish farm.

Figure (5-9a) shows the fish ponds around the lake area

Figure (5-9b): shows the pumping stations of Barseek fish farm.

Effect of lake Vegetation on Water Quality and Lake Fisheries

This vegetation has two main impacts on the lake and its fishery. There is a tendency to block water flows and to present an obstacle to movements of the fishermen. In addition to the physical problem of passage through the weeds, they may have an effect on water quality by stagnating water volumes and leading to extremely anoxic and foul conditions. Among the biological effects of the weed growth is the suppression of plankton development by the uptake and locking away of the available nutrients. At the same time there is the provision of a vastly increased surface area for the establishment of epiphytic organisms and thus enhancing the available food supplies for many of the fish species.

It is difficult to judge to what extent the growth of the reeds and the water hyacinth mats affect the flow of water through the lake, as there is no evidence of backing-up in the drainage channels except during the periods where the water hyacinths nourishes. There may be an effect on the circulation of water throughout the lake area, resulting in the patchy changes in water quality observed by the consultants. Physical obstruction of the fishermen's access channel does occur and can arise from the movement of relatively small amounts of water hyacinth. The blocking by reeds or submersed aquatics does not present a problem that cannot be dealt with by the fishermen themselves.

The effect of weed growth on the production of fish, and the consequent yield to the fishery, is complex. The fact that many areas of the lake are inaccessible to the fishermen, provides a safe refuge and breeding area for many of the fish species. These inaccessible areas act as a conservation measure and will prevent overfishing. Such an "automatic" control over fishing activity is therefore an advantage for the long-term sustained production of the fishery.

The large expanses of submersed aquatic weeds provide a well oxygenated, food-rich habitat, but, on the other hand, at night may provide conditions low in dissolved oxygen and therefore may have a negative effect on fish production. If the submersed weeds are eliminated, there will be an increase in dissolved nutrients available to the phytoplankton populations. The shading effect of an increase in these populations will aid in retarding the growth of the sub-aquatic weeds and the underwater parts of the emergent species. The development of such phytoplankton populations would provide an increase in the food supply for the tilapias and the mullets (especially in their smaller stages). Any control of the weeds on the lake should be considered from the separate aspects of the biological effect on the fishery and the improvement in the work and life of the fishermen.

The lake receives a huge amount of drainage water (5.2×10^6 m^3/day) from khairy, Boussily Edko and Barseek drains along the southern and eastern sides. Figure (5-10) shows the main three physical components of the lake system. Seawater primarily affects the western side of the lake near the outlet. After construction of the Aswan High Dam, the annual drainage to the lake has increased. This has caused an increase of the level of the lake and induced flow from the lake into the sea such that the lake became less influenced by salt water from the sea. This initiated east-west flow of water in the lake and caused a slight elevation of its water above the sea level. In the northern part of the lake area the water body is traversed by the international coastal way bridge dividing the lake area into two different parts regarding water vegetation which varies from upstream and downstream of the road. Most of the floating vegetation and fishing activities are on the upstream side of the road bridge. The downstream side is considered very shallow compared with the upstream part except at the outlet channel to the sea.

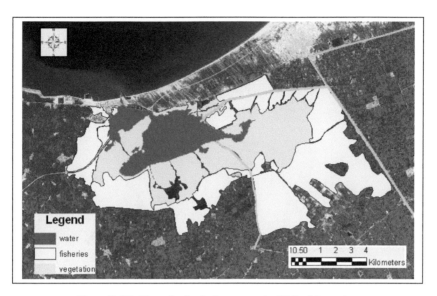

Figure (5-10): The main physical components of Edko lake system

5.5.3. Climate of the Study Area

The climate of Egypt is governed mainly by its location in the north-eastern part of Africa on the margin of the largest desert in the world. Its latitudinal position, between 22° and 32° N place it firmly in the sub-tropical dry belt, although conditions on its northern coast are ameliorated by the presence of the Mediterranean Sea. The climate is considered arid to semi-arid, throughout most of the year the hot, dry tropical continental air masses dominate, but during the winter period air masses of both tropical maritime and polar maritime origin make brief incursions into Egypt from the north, frequently bringing rain with them.

The average mean maximum temperatures in the delta region and coastal area varies from 18-24 C° in winter and 37-42 C° in summer, mean minimum temperatures vary from 4-9 C° in winter and from 16-17 C° in summer. The average annual rainfall on the catchment and lake area is less than 125 mm/year during winter rainy months (January, February). The evaporation rate is high and reaches around 2000 mm/year. Maximum evapotranspiration is recorded in summer (June-August) with a rate of 6.88 mm/day. According to collected data on the study area the wind prevailing direction in this coastal region is NNW with an average speed of 5 m/sec.

5.5.4. Hydrology and Drainage

All the drainage water discharging from the catchment area including the agricultural drainage water and the discharges from fisheries ends up int the Lake main water body. The water enters the lake through the two main drains Edko and Barseek. The lake receives an average amount of 50.2 m3/sec. drainage water from Edko main drain and an amount of 11 m3/sec from Barseek Drain. Therefore, the lake receives an average total daily discharge of 60.3 m^{3}/sec. Figure (5-11) shows the two main drains contributing to Lake Edko.

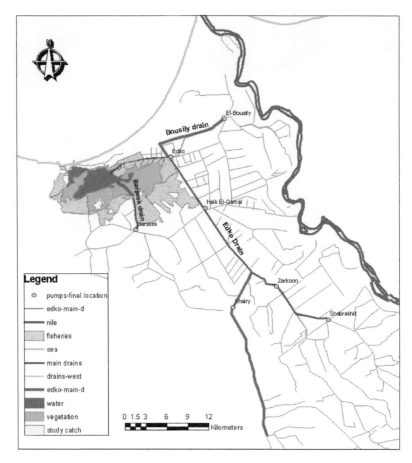

Figure (5-11): Main drains discharging into Edko Lake

5.5.5. Water Uses, Land Uses and Human Activities

Edko watershed is a part of the western delta region; it is an agricultural watershed, the main water use being agricultural, in addition to municipal and industrial water uses. The amount of water used in the cultivated area of the watershed is estimated to be 1793 km^2 (427000 feddans). The main land use in the watershed is agriculture, with some land devoted to villages and small towns, some industrial plants and fish ponds around the lake area. Also the land use reflects the main human activities in the watershed, which are mainly agricultural activities, some industrial activities like oils and textiles (factories exists in the upstream area of the catchment), and aquaculture and fishing activities inside and around the lake area. In figure (5-12) we see a map that illustrates the different land uses in the watershed.

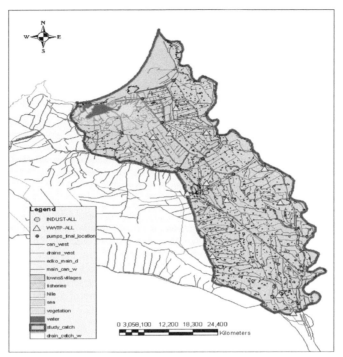

Figure (5-12): Land Uses Edko Catchment-Lake System

5.5.6. Water Quality Problems

The problem of surface water quality in the selected watershed includes the pollution of the drainage network and the northern lake Edko. As part of the Nile Delta, the watershed suffers from different pollution sources due to different human activities. The drainage network is subjected to point sources in the form of industrial waste discharges and secondary wastewater treatment plant disposals into the drainage network at different locations. Non-point source pollution occurs due to the fertilizers and pesticides applied in agriculture, in addition to domestic wastewater disposal and untreated waste disposal into the drainage network. Drainage water reuse is extensively carried out in the watershed through the presence of four reuse pumping stations. One of these reuse pumping stations discharges reuse water into Mahmoudia canal, which is the main source of drinking water for Alexandria city.

The northern lake Edko could be defined as a coastal lagoon. It suffers from excessive vegetation growth and eutrophication due to the excess of nutrients discharged into the lake, in addition to other sources of pollution which are mainly: untreated waste water, industrial waste water from different industries and salt water entering from Bugaz channel connected to the sea which is full of industrial untreated wastes dumped to the Mediterranean sea during tidal periods. The main pollution problem in the lake water in addition to the nutrients is the contamination of the water with fecal bacteria. According to El-Shennawy et al (2000) it was found that there is a prevalence of indicator bacteria including coliform bacteria, e-coli and fecal streptococci as well as some other pathogens. This is considered a threat to the human health of citizens living in the lake area and depending on fishing and aquaculture for their daily living.

Water quality data from monitoring points in the catchment is collected on a monthly basis; raw data for water quality parameters is available for the period from 1999 -2005. For the lake there is no fixed monitoring network but single studies were conducted during 1991, 2000, 2001, 2002, dealing with different water quality parameters such as some hydrochemical parameters and some biological parameters in the lake water, and the study of occurrence of humic substances in lake sediments.

5.6. DATA SOURCES AND ANALYSIS

Assessing and managing water quality requires a huge amount of data to be collected. In order to develop the water quality management information system, different types and categories of data are needed to develop such a system. The following section identifies all the available and collected data that are needed as inputs for developing the system components. The different data sets were collected and analysed for the study watershed including the catchment area and the shallow lake. The different categories of data were identified based on the required data needed to be included in the geo-database, which is the main core of the information system.

Three different categories of data are identified here: *Digital maps and GIS layers, different attribute data collected from available sources and from field work, satellite images and remote sensing data*. The collected data is divided into two sets: *Secondary and primary data sets*. The *Secondary data sets* are the data gathered from different sources such as databases, reports and documents and accounts of previous work done. Most secondary data were obtained from reports and databases from governmental institutions and agencies. Where there was a gap in the secondary data, *primary data set* collection was undertaken mainly through field work. The objective of the primary data survey and collection was to fill in the data gaps in the secondary data sets.

5.6.1. Data Inventory

The different types of data that is needed from both the secondary and primary data collection, to feed the surface water quality geo-database includes: Flow rates in main drains, pumping station data, water levels, water quality data, tidal variations, meteorological data (wind - evaporation rates - solar radiation – humidity - temperatures), ground water boreholes, demographic data and population rates, geometric data and cross sections of drains, digital maps and GIS layers of the catchment and the lake areas, bathymetric maps of the lake with different resolutions, satellite images of different types and resolutions, water samples at certain locations in the catchment and lake, water quality field spectrometry readings. The collected data were grouped according to their properties into data sets that are shown in table (5-1) which shows the first level of data collection (*Secondary collected data*).

Secondary Data Sets

The secondary data sets include all types of needed data for the development of the water quality management information system geo-database. The following section gives an explanation of different types of collected data.

A.   Hydrodynamic data sets

This data set includes the discharges and flow rates from the main drains in the drainage catchment. The flow rates were given at each of the pumping station existing on the main and secondary drains in the catchment. A number of 6 pumping stations exists on the main drain Edko and discharges water to the main drain, and one pumping station exists on the other main drain in the catchment, barseek drain. Average hourly discharges of these pumping stations are recorded and used in the development of the hydrodynamic models components. Table (5-2) shows the maximum flows from the six pumping stations to Edko drain along with the permissible maximum water level at each station.

Table (5-1): First level of data collection (Secondary collected data)

No.	Data type	Data source	Details and remarks
		Exisitng Secondary Collected data	
		Hydrodynamic Data Sets	
1	Flow rates in main drains	DRI Database	data from 2003-2006
2	Pump stations flow data	study report HRI	average flow rates (60 min time step)
3	Tidal water level varaition	developed for study area from x-tide software	March -Sept. 2005 (30 min time step)
4	Ground water boreholes data	RIGWA Database	average flow data
		Meteorological Data	
1	Wind direction and velocity	EMA and NREA	monthly average data for study area
2	Temprature	EMA	monthly average data for study area
3	Evaporation rates	EMA	monthly average data for study area
4	Solar radiation	EMA	monthly average data for study area
5	Relative humidity	EMA	monthly average data for study area
		Water Quality Data Sets	
1	Physicochemical parameters	DRI and CIEQM Database	data from 2002-2006 monthly records
2	Major cations	DRI and CIEQM Database	data from 2002-2006 monthly records
3	Major anions	DRI and CIEQM Database	data from 2002-2006 monthly records
4	Trace metals	DRI and CIEQM Database	data from 2002-2006 monthly records
5	Microbiological parameters	DRI and CIEQM Database	data from 2002-2006 monthly records
		Geometric Data and Cross Sections	
1	Main drain Edko cross sections	HRI	142 cross sections with 200m spacing
		Digital Maps and GIS Layers	
1	Digital maps of catchment and lake	developed based on topographic survey maps of scale 1:50000	digital maps are developed using Arc-GIS
2	GIS layers of the study area	developed and projected in UTM coordinates	layers are developed using Arc-GIS
		Satellite Images	
1	ASTER	ITC Remote Sensing Laboratory	Images from 2004-2005
2	MODIS	ITC Remote Sensing Laboratory	Time series set of images (for 2005-2006)
3	SPOT	Purchased from SPOT website	one scene on 7th of July, 2006
4	Landsat TM	ITC Remote Sensing Laboratory	group of images from 1970-2006

Table (5-2): Maximum permissible flows and Water Levels for the Pumping Stations on Edko Drain

Pump Station	No. of Working Units	Flow per Unit (m3/s)	Total Flow (m3/s)	Maximum Allowable Water level (m)	Area Served (feddan)
Shubrakeet	3	6	18	1.8	25,000
Zarkoon	3	5	15	1.6	41,000
El Khairy	3	6	18	1.5	75,000
Halk El Gamal	4	5	20	1.1	56,100
Edko Drainage	2	3.5	7	0.8	18,000
Boseili	2	7.5	15	1.45	25,000
Total			93		240,000

Note: I feddan = 4200 m^2

The locations of the pumping stations are shown in the catchment layout given in Figure (5-13). These pumping stations are also the fixed water quality monitoring stations in the catchment. At the entrance of the main drain Edko to the lake, there is a fixed measuring point of the flow to the lake; monthly average records are available at this point from the years 2000-2006.

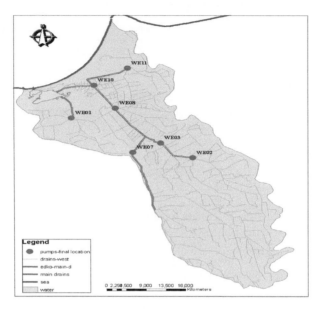

Figure (5-13): locations of pumping stations

Tidal Water Levels
The other main component in the hydrodynamic data set is the *Tidal Water Levels* variations. No permanent station measures tidal flow at the lake exit channel to the sea. The Coastal Research Institute performs some spot checks but does not keep systematic records. To overcome the lack of tidal time series, which is a crucial input to the hydrodynamic model, a detailed survey in the literature was done to understand the tidal properties at the location of the exit channel to the Mediterranean Sea. Based on this information the WX-Tide software package was used to predict the tide and currents according to the geographical location of the study area for a duration of six months.

Groundwater
Another important component of the hydrodynamic data sets is the information on the ground water flow rates and pezometric heads in the western delta region around the lake area. That information is needed to interpret for the water losses from the lake to the ground water table. The groundwater aquifer underlying the Nile valley and Delta is recharged by seepage losses from the Nile, the irrigation canals and drains, and deep percolation of water from irrigated lands. The total available storage of the Nile aquifer is estimated at about 500 BCM but the maximum renewable amount (the aquifer safe yield) is around 7.5 BCM. The existing rate of groundwater abstraction in the Valley and Delta regions is about 4.8 BCM/year, which is still below the potential safe yield of the aquifer.

Field and laboratory experiments were carried out to determine the hydraulic parameters for the Nile Delta aquifer. According to Sherif et al, (2002) and based on field data, an isotropic hydraulic conductivity of 100.0 m/day and a storativity of about 10^{-4} to 10^{-3} were considered representative of the regional values of the aquifer. Farid (1980) reported different values for hydraulic conductivity and storativity at various locations. The hydraulic conductivity of the aquifer decreased toward the south and west. An effective porosity of 0.3 was considered to represent the aquifer medium. The free water table is measured at various locations. When this information is missing it is generally assumed to be 1.0 m below the ground surface. The piezometric head is measured periodically through an intensive-monitoring network. Records for water levels and piezometric heads are available in the database of the Groundwater Research Institute (GWRI), National Water Research Center (NWRC) of Egypt.

B. Meteorological Datasets

Several meteorological datasets are needed as basic inputs into the hydrodynamic and water quality models. The data sets needed for this purpose are: *wind directions and velocities, temperature, relative humidity, solar radiation and evaporation rates.* Due to the important effect of wind on the hydrodynamics of the lake system, the hourly wind directions and velocities are collected. The detailed analysis of wind data in the study catchment is shown in the chapter hydrodynamic modelling the wind data sets were collected from meteorological station of Alexandria sea port, which is the nearest station to the study area location. The data sets cover the period of two years (2005-2006) for identification of average wind characteristics in the study area. For the hydrodynamic model the data from January up to End of July, 2006 is used. The air temperature was also collected for the same period of two years. Relative humidity, average solar radiation and evaporation rates monthly average data were collected from the Egyptian Meteorological Authority (EMA) for the years 2005-2006.

C. Water Quality Datasets

The water quality datasets for the monitoring stations in the drainage catchment were collected from the database of the Drainage research Institute (DRI) and the Central Laboratories of Environmental Quality Monitoring (CLEQM) of the NWRC of Egypt. The data sets include average monthly records from the year 2000 until 2006. The available data on water quality of the drainage network include physicochemical, major anions and cations, trace metals and microbiological parameters. No water quality data or measures were available for the lake area and that was a main reason for the need of field work to collect water samples from the lake for the implementation of research work. A detailed map of the different locations of the monitoring stations is shown in the coming section.

D. Digital Maps and GIS layers

The topographic base maps covering the watershed area were collected, having a scale of 1:50000, and the main needed GIS layers were digitized from these maps. The GIS layers are projected according to the UTM projection coordinates for the study area (UTM, WGS-1984, Zone: North36), this is the same projection for all the spatial data used in this research work. The basic GIS layers include the following:

• **Western Delta boundaries**	• **Cities and villages**
• **Mediterranean Sea**	• Lake boundaries
• **Nile Rosetta Branch**	• Vegetation areas
• **Main Study area boundaries**	• Fisheries areas
• **Irrigation network**	• Pumping stations
• **Drainage network**	• Wastewater treatment plants
• **Agricultural lands**	• Industries

Other specific GIS layers are developed from processing of collected data through field work missions. These layers include the following:

• **Lake Bathymetry raw data**	• **Spectral measurements locations**
• **Water depth developed grid layers**	• **Water levels measurements**
• **Water quality sampling locations**	• **Measured discharges locations**

Another set of GIS layers is developed from the water quality modelling outputs and linked spatially to measurement points locations for models calibration and verification purposes. Also processed satellite images outputs are liked spatially to these locations. More details on the GIS layers are explained in the section of the geo-database.

E. Geometric Data and Cross Sections of Main Drain Edko

This data set is of specific importance because the development of the 1D hydrodynamic catchment and drainage network model is developed based on it. This dataset is collected by HRI earlier in 2003 for a survey study conducted on Edko drain. The data set includes 142 geometric cross sections of the main drain Edko, showing the detailed geometry of the main drain. In addition to the geometry of the drain the water levels at each cross section is measured and used in the modelling work.

F. Satellite Images

This research work mainly depends of the use of satellite images in different stages of work development. Therefore, a variety in number and types of satellite images are used. A number of ASTER images are collected for the detailed delineation of watershed components, water bodies, vegetation and fisheries. Three images of Landsat TM were selected to identify the spatial changes in the lake area and to support the catchment delineation task. The major set of images used is the MODIS time series (MOD09GQK) that extend within the modelling period of the lake. This set is used for both the qualitative and quantitative calibration of the water quality models of the lake. A SPOT-5 image (quarter scene) for the lake is used for the detailed quantitative calibration of the eutrophication screening model. Another important source of images was ENVISAT-MERIS image taken at the same date of the SPOT Scene. Table (5-3) shows the used satellite images, acquisition dates and the source of images. Detailed characteristics of images are presented in chapter 8.

Table (5-3): Selected satellite images

No.	Sensor	Acquisition Date	Source	Remarks
1	ASTER	October 19, 2004	ITC, The Netherlands	Downloaded via ITC
2	Landsat TM	1972-1984-2002	ITC, The Netherlands	Downloaded via ITC
3	MODIS (MOD09GKQ)	(Jan. - July 2006)	ITC, The Netherlands	17 images (see chapter 8)
4	SPOT-5 (HRG1)	July 10, 2006	www.spotimage.com	Purchased
5	ENVISAT-MERIS	July 10, 2006	ITC - ESA-TIGERII CB Project	Downloaded via ESA-EOLISA software

5.6.2. Water Quality Monitoring Locations within the Study Area

The DRI has maintained a water quality-monitoring network for some 18 years. The main objective of DRIs monitoring effort is to provide MWRI with information on the availability of drainage water for reuse. Another objective is to identify changes in\ drain water quality caused by municipal, industrial and agricultural wastewater discharges to the drains. The DRI monitoring network includes 169 stations in the Delta and Fayoum to monitor ambient water quality, primarily in drains. The DRI Nile Delta monitoring network is shown on Figure (5-14) Sampling takes place monthly. A breakdown of the total number of sampling points is as follows:

Fayoum Area – 8 points on drains and 4 on canals
West Delta – 39 points on drains and 7 on canals
Middle Delta – 41 points on drains and 10 on canals
East Delta – 43 points on drains and 16 on canals

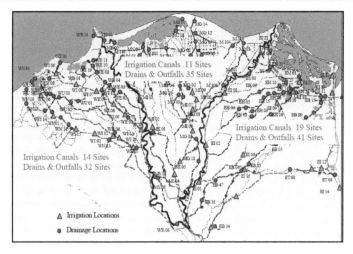

Figure (5-14): Water Quality monitoring Network in the Nile Delta Region

Within the drainage catchment of Edko, as mentioned earlier there exist 13 water quality monitoring locations 12 of which are related to Edko Drain and one on Barseek Drain. Table (5-4) shows the pumping stations and other monitoring locations in Edko catchment. It has to be mentioned that there is no monitoring network for coastal shallow lakes in Egypt that is working on regular basis for the entire coastal lakes region. There are no archived or historical records of water levels, or water quality measurements inside the lakes. The existing measurements are based only on scattered projects, and pilot case studies (C-CORE, 2007). Since there was no sufficient data to conduct a detailed study on the hydrodynamics or water quality of Edko Lake, it was essential to conduct field work for acquisition of all needed data and parameters for models development.

Table (5-4): shows the pumping stations and other monitoring locations in Edko catchment
(Source: DRI-NWRC)

Location Code	Location Name	Type	Measurement Location
WE01	Etay El Barud P.S.	Reuse P.S.	before entering the irrigation system.
WE02	Shubrakhit P.S.	Drainage P.S.	before entering the beginning of Edko drainage system.
WE03	Zarqun P.S.	Drainage P.S.	before entering Edko drainage system.
WE04	Edko Irrigation P.S.	Reuse P.S.	before mixing with the irrigation system of Mahm. Canal.
WE05	Dilingat P.S.	Reuse P.S.	before mixing with the irrigation system.
WE06	Khandak El Gharby P.S.	Reuse P.S.	before mixing with the irrigation system.
WE07	Khairy P.S.	Drainage P.S.	before entering the main drainage system of Edko drain.
WE08	Halq El Gamal P.S.	Drainage P.S.	before entering Edko drainage system.
WE10	Edko Drainage P.S.	Drainage P.S.	before entering Edko drainage system.
WE11	Bosseily P.S.	Drainage P.S.	before entering Edko drainage system.
WE13	Diab Bridge	Open Location	at Edko drain downstream Halq El Gamal P.S.
WE19	Edko Outfall	Open Location	before entering Edko lake.
WE20	Khairy drain at Agriculture road	Open Location	Kairy drain after mixing with Damanhour seawage water.
WE21	Edko drain after Khairy outfall	Open Location	Edko main drain after mixing with Khairy drainage water

5.6.3. Field work and Data Collection

Primary Data Sets

The primary data sets are considered the crucial and complementing component to the secondary datasets. This data was possible to be collected through on-site field work because it was not available in databases or archives like other types of secondary sets. Also some measurements are study specific and need to be collected during field work. This field work is considered an essential and important tool in this research. The physically based models developed in this study are based on field measurements and investigations of different parameters used for modelling, for both the hydrodynamics and water quality. Four field work surveys were conducted at different times during the period of research. Each specific field mission has its own objectives and type of data needed for conducting the work. The details of files work is given in given below.

Field Work (1)

This field work mission was conducted in *May 2005*, as a primary and main investigation of the catchment and lake area and for collecting basic needed data to develop the modelling tools. A strategy for field work and data collection was planned. The following section summarises the field work plan;

Required Data Collection
Data collection and fieldwork in the study area is divided into two parts; the catchment fieldwork and the lake field work. Figure (5-15) shows the sampling sites location within lake.
- **Catchment field work**

The fieldwork in the catchment will include data collection along the main drain Edko and at the points of pumping stations within the study catchment The required data include:
  - o Locations of main pumping stations and bridges along main drain Edko (X,Y coordinates, using GPS in UTM format)
  - o Water quality samples at the locations of pumping stations and bridges along the main drain Edko
- **Lake Edko field work**

For the lake survey the following data is required:
  - o A number of **five points are selected along the lake area**. Total station survey for the lake area was conducted using these points to identify levels, and was used as reference points for other measurements in the lake
  - o Taking measurements inside the lake using GPS eco-sounder to identify water depth at different possible locations that were accessible.
  - o Taking water samples for water quality analysis of different parameters and specifically physicochemical, biological and CHL-aorophyll a
  - o Taking water quality samples inside exit channel of the lake to the sea
  - o At the same locations of water sampling, a number of sediment samples were collected.
  - o Detailed identification of the vegetation types inside the lake through photos and sampling of vegetation.
  - o Measuring water discharge from Edko drain, through measuring water velocity cross sections at a fixed location along drain outlet via current meter and at the exit point of the lake to the sea channel.

Equipment
The equipment used to conduct the fieldwork is the hydrographic survey equipment (Total station, GPS, Eco-sounder, Current meter and bottom sediment sampler) and the water quality sampling

equipment (sampling bottles, filter paper, field kits for measuring PH, temperature and EC). Figure (5-15) shows the catchment lake area and the water quality sampling locations.

The first important item in this field work was to have a complete hydrographic survey of the lake water body, in order to have a bathymetry map of the lake. This hydrographic survey was conducted using field GPS-Eco-sounder survey equipment. Another important item in this field work was measuring water levels at different locations within the catchment and the lake area and for this the total station survey and levelling was required. Collection of water samples at several locations along Edko main drain, at pumping stations and inside the lake for physicochemical and biological analysis was a main target for preliminary assessment of water quality conditions. A total number of 19 samples were taken, ten samples out of this set were also taken for analysis of CHL-a. In addition, a number of four soil samples were collected from the lake bottom for general interpretation of the soil type and concentrations of chemicals and trace metals within the sediments. Knowing the importance of the effect of submerged vegetation a sample from *Potamogeton pectinatis* was taken from the lake for lab analysis of the chemical constituents in the plant tissues. Two velocity cross sections were measured at the exit of the drain to the lake and at the exit of the lake to the sea channel.

Water quality samples were collected preserved and sent forms the sampling location at Edko Lake to the Central laboratories for Environmental Quality Monitoring (CLEQM) at Delta barrages for physicochemical and biological analysis. The CHL-a samples were collected and sent directly for analysis to the Central Laboratories of the National Institute of Oceanography & Fisheries (NIOF) in Alexandria. The results of water quality analysis from filed data collection for some of the basic parameters are shown in table (5-5).

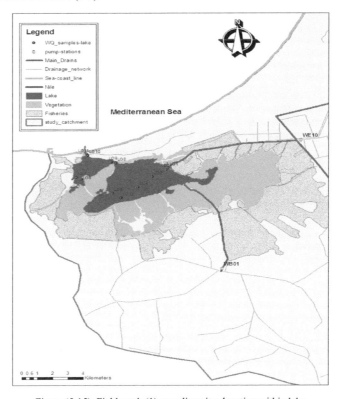

Figure (5-15): Field work (1) sampling sites location within lake.

Table (5-5): Analysis results of water quality samples from lake water body (2005)

No.	Sample ID	pH mg/l	EC mmhos/ cm	TDS mg/l	DO mg/l	NH4 mg/l	NO3 mg/l	PO4 mg/l	BOD mg/l	COD mg/l	Salinity g/l	Chl_a µg/l
1	LO1	7.53	1.71	1095	1.4	2.7	18.53	0.59	12	19.2	0.7	42.87
2	LO2	7.57	1.7	1085	1.4	3.5	20	0.65	11	16.5	0.7	41.85
3	LO3	7.39	1.63	1047	5.8	1.6	18.7	0.69	19	30.4	0.7	47.83
4	LO4	8.7	2.85	1823	8.8	2.8	26.88	0.92	14	22.4	1.4	68.16
5	LO5	8.95	2.6	1665	8.3	0.73	9.64	0.48	18	30.6	1.3	61.84
6	LO6	8.89	2.37	1515	11.8	0.52	11.69	0.21	18	31.4	1.1	71.2
7	LO7	8.99	2.33	1489	13.2	0.68	15.53	0.56	19	32.3	1.1	75.86
8	LO8	8.85	2.21	1415	13.2	1.14	9.7	0.34	19	34.2	1	70.73
9	LO9	8.67	1.82	1168	14.6	1.1	15.3	0.7	19	30.2	0.8	74.99
10	LO10	8.78	2.04	1306	14.12	0.85	11.7	0.61	16	27.2	0.9	76.69
11	LO11	8.51	1.89	1214	10	1.65	24	0.81	16	25.6	0.8	50.9
12	LO12	9.03	2.44	1564	10.93	0.68	7.1	0.23	17	28.9	1.2	49.74
Average		8.48833	2.1325	1365.5	9.4625	1.49583	15.731	0.5658	16.5	27.4083	0.975	61.055

Field Work (2)

This field work was conducted in *January 2006*. It was specifically to collect water samples from Fish ponds in several locations around the lake area including the Barseek pumping station location. Due to the difficult access to the locations of fish ponds it was possible to collect a number of 5 samples to represent different areas around the lake. The main target of this sampling is to check the concentration of nutrients within the ponds and compare it with the concentration within the lake and within the drains, for initial estimate of the level of nutrients discharged from fisheries. The samples analysis showed high levels of salinity that ranges from 1.9 mmhos/cm to 5mmhos/cm in some locations. Also high concentrations of COD are measured, that ranges from 38 to 288 mg/l in some ponds. For the nutrients ranges of Nitrates (NO3) were from 13.5 to 25.4 mg/l. It is noticed that the water in the fish ponds tends to be Alkaline, the pH measured values ranged from 8 to 8.9, and this is also shown from the alkalinity measurements that ranges from 290 to 532.

Field Work (3) (Major Data Collection)

This field work mission was conducted in *June 2006* and it is considered the main mission for water quality sampling and data collection for the calibration procedures of the models. This data set was essential for the water quality models calibration and remote sensing application for models calibration. Figure (5-16) shows the water quality sampling locations inside the lake. The following data sets were measured;

- A set of 21 water quality samples covering the lake area and the end of main drain Edko. Six only of these points were measured for CHL-a due to time and facilities limitation in field. The whole set was sampled for physicochemical and biological analysis.
- For the same 21 locations within the lake water body spectral reflectance measurements were taken using hand-held (ASD field-spectrometer). At each location a number of 9 spectral measurements were taken for data confirmation and selection of most suitable measurement. *(Details on field spectral measurements is presented in the next section)*.
- In addition to the 21 spectral measurements of water, 3 additional spectral measurements were taken about other objects, two of which are from mixed and submerged vegetation in the lake water and the third from sand bars on the lake shore. The vegetation samples were needed to be able to define how much of the lake pixels are affected by the lake vegetation.

- Two velocity cross sections were measured at the fixed discharge measuring point in Edko main drain near exit to lake, and near the bridge on sea exit channel. These sections were required for checking the inflow and outflow from the lake water body.

Field Work (4)

This field work was conducted in **November 2007,** and the main need was to measure velocity cross sections at the same locations or earlier measurements, for comparison and check of lake water balance. The other important measurement was to check the tidal variation of water levels at the sea exit channel, water levels recording was done for continuous 12 hours inside the exit channel.

5.6.4. Description and Interpretation of Remote Sensing Data Sets

In situ Water Quality Reference Data

The in situ water sampling and quality measurements were carried out in order to get the true water quality parameters values of Edko lake. As shown in Figure (5-16) water samples were collected at selected 26 sampling locations for TSM, Secchi disk transparency, pH, turbidity, depth water temperature, dissolved oxygen (DO), orthophosphates as phosphorus (PO4), total phosphorus (TP), Nitrogen compounds as ammonia (NH4), nitrate (NO3) and nitrogen N, 5 days biochemical oxygen demand (BOD) and chemical oxygen demand (COD).The samples were taken on three days field work from June 27th till June 29th.. The main parameters of concern for remote sensing in this study are the CHL-aorophyll a, TSM, turbidity and Secchi depth.

CHL-aorophyll *a* values were determined by performing analyses on water samples collected at six selected locations. CHL-aorophyll pigments were extracted in aqueous acetone (90% acetone:10% water) and analyzed spectrophotometrically (at the Research Center for Oceanography, Alexandria). Total Suspended Matter (TSM) was defined for all 26 sampling locations. TSM samples were analyzed in the laboratory using filtration of a known amount of water through a pre-washed, pre-dried at (103-105 °C), pre-weighed ($\sim \pm 0.5$ mg) filter - rinse, dry and reweigh to calculate concentrations in mg/l. Secchi disk depth readings were done at each sampling location. Secchi disk depth is used as an important indicator of trophic state and overall clarity of the water. Field turbidity measurements are made with Submersible portable turbidimeter.

Field Spectrometer

Spectral reflectance measurements were taken with an (ASD) Field Spec Pro-handheld field-spectrometer which collected reflectance from 350 nm to 2500 nm. A white reference disk was used to measure incoming radiance for absolute reflectance calculation. The sensor (with a conical field of view of 25°) was attached to a 1 meter long metal bar and extended away from the boat so that shadow and surface effects from the boat didn't affect the readings. The spectrometer was held roughly 0.6 meters above the surface of the water with a collection time of 25 seconds. Reflectance readings were gathered at the same locations that the water samples were taken. Days that reflectance were collected were mostly sunny days so that sky reflectance was minimized. At each sampling location, after a refrence panel measurement, a number of nine spectrums were taken with a white reference as the tenth spectrum. A total number of 26 spectral sampling sets were measured during three days of field work. Spectral measurements of vegetation from both the floating vegetation (water hyacinths) and the submerged vegetation (potamogeton Pectinatus) were also taken. These measurements were considered of importance due to the presence of these two types in the lake water and their significant effect on the signal of spectral measurements. Figure (5-17) shows the spectral samples measured at the different selected locations inside the lake.

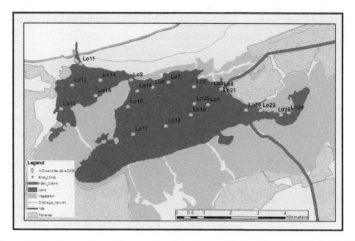

Figure (5-16): The water quality sampling locations inside the lake, field Work (3), 2006.

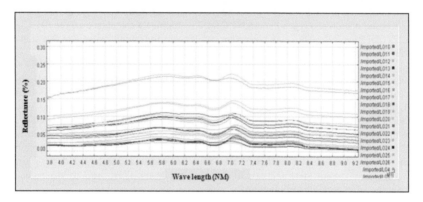

Figure (5-17): Spectrometer measured samples (Remote Sensing Reflectance) from Edko Lake (June, 2006)

The general analysis and interpretation of water quality within both the catchment and the lake based on the secondary and the primary data sets are given in the next section. Seasonal trends in pollution loads and variation of pollution concentration within the drains and lake water body are discussed in the coming section. This section also will include highlights on the earlier water quality research done on either the Edko drainage catchment or the lake.

5.6.5. Water Quality Assessment

In this section preliminary investigation of the current situation of water quality in the catchment and lake are given. This interpretation of existing conditions is based on the collected data from both primary field work and secondary archived data. First, it is important to investigate the water quantities that are flowing into the main drain from the catchment system and consequently into the lake. The main drain Edko receives all drainage water from the catchment through the mentioned six pumping stations. Figure (5-18) shows the distribution of discharge from different stations into the main drain Edko.

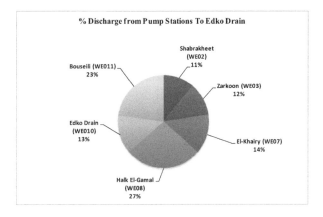

Figure (5-18): Percent discharge from Pumping stations to main drain Edko.

The other source of discharge into the lake is from Barseek drain, Figure (5-19) shows the average monthly discharges from both drains into Lake Edko.

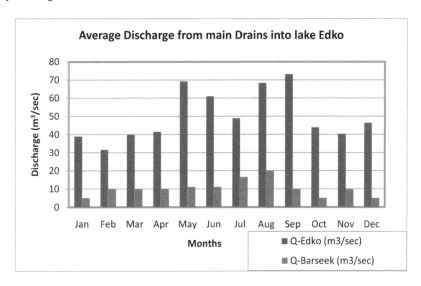

Figure (5-19): Average monthly discharges from drains into Lake Edko.

Water Quality Analysis from Catchment Monitoring Locations

In the drainage catchment monitoring of water quality is conducted at several fixed locations as mentioned earlier, for the main Drain Edko. At the six pumping stations locations water quality sampling is recorded since 2001. The following section shows the main selected water quality parameters recorded at specific locations of pumping stations along the drains (WE02, WE03, WE07, WE08, WE010, WE011,WB01). Figure (5-19) shows the selected locations along the main drains Edko and Barseek. The selected water quality parameters are; (pH, COD, BOD, TSS, TDS, DO, Temp., Turbidity, Salinity, TP, OPO4, NH4, TN, NO3). The sampling locations are divided into three zones; upstream zone including stations (WE02, WE03, WE07, WE08), downstream Zone including (WE010, WE011) and Barseek Zone (WB01).

For these sampling locations, data were collected from the year 2001 up to 2006 on monthly basis; the preliminary analysis for these data sets was carried statistically to show the seasonal and spatial trends at different locations within the drainage catchment. Figure (5-20) the grouping of the measuring location within the catchment area according to sub-drainage catchments. The monthly data shows seasonal variation for the different measured parameters.

Upstream Zone Edko Drain
This zone receives drainage water from four pumping stations including domestic, agricultural and industrial wastewater according to the different land use activities in this area. This zone receives 37% of the drainage water entering the Lake from Edko drain.

Downstream Zone Edko Drain
This zone receives mainly agricultural drainage water and domestic waste water from three pumping stations, and it receives 63% of the total drainage water entering the Lake from Edko drain

Barseek Zone
Barseek zone includes one pumping station that receives drainage water from southern part of the lake this zone is most associated with drainage water from fish farms and agricultural areas.

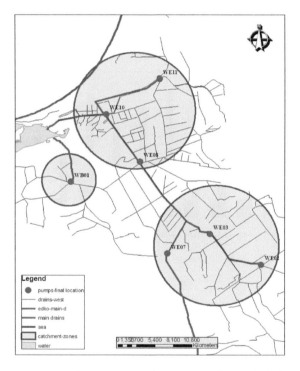

Figure (5-20): grouped water quality measurement locations according to sub-drainage catchments

The following section gives an over view of the concentration of some main parameters at the different catchment zones.

At Location WEO2 (Upstream Zone)

Figure (5-21) shows an average monthly salinity of 1.16 g/l for the years 2001-2005 in the drainage water entering Edko drain from Zarkoun pumping station (WEO2).

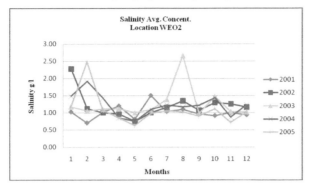

Figure (5-21): average salinity US Zone

For TP and TN and TSS the yearly concentration values vary between avg.max of 1.11 mg/l and 0.27 mg/l for TP, avg. max 26.35mg/l and avg. min 4.07 mg/l for TN, avg. max 100 mg/l and avg. min 30 mg/l for TSS. Figures (5-22), (5-23) and (5-24), shows these values for the years from 2001-2005.

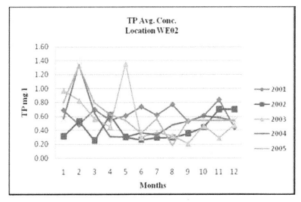

Figure (5-22): Average TP US Zone, Location WE02

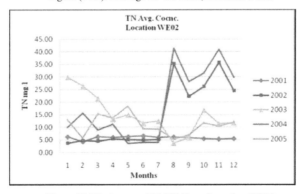

Figure (5-23): Average TN, US Zone, Location WE02

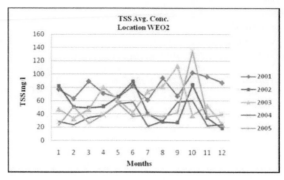

Figure (5-24): Average TSS, US Zone, Location WE02

At Location WE10 (Downstream Zone)

Figure (5-25) shows that the average monthly salinity entering downstream zone of Edko drain is increased to 2.1g/l with a maximum avg. of 4.6 and a minimum avg. of 1.2 g/l.

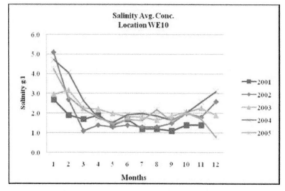

Figure (5-25): average salinity DS Zone

The average yearly measurements for TN at that location reveal a very high concentration of TN during the year 2004 with an average of 43.57 mg/l which exceeds the yearly averages from 20-30%. These high concentrations at the location WE10 which is the confluence point of Edko drain with Buosiely, are measured at Edko bridge, 8 km from lake entrance. The reason to this high concentration could be due to discharge of untreated waste water or drainage water with high ammonia fertilizers. Figure (5-26) shows these extreme events during 2005.

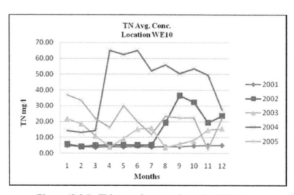

Figure (5-26): TN Avg. Conc. At Location WE10

The average maximum values of TP for five years are 0.54 mg/l and average minimum values of 0.3 mg/l. Figure (5-27) shows higher values for TP at the exit point to the lake during the year 2005 of around an average value of 0.68mg/l.

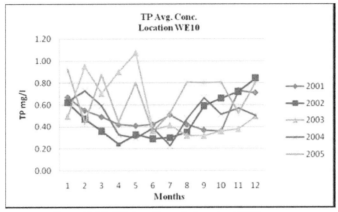

Figure (5-27): TP Avg. Conc. At Location WE10

It is noticed from Figure (5-28) that the values of TSS are increased almost by double values during the years 2003, 2004, and 2005.

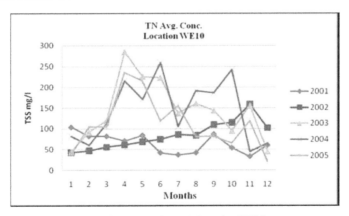

Figure (5-28): TP Avg. Conc. At Location WE10

To summarise the situation of the water quality concentrations discharged from the catchment into the lake, the figures shown above illustrates that the concentrations of pollutants increases with the direction of flow as they accumulate from different points of inflow along the main drain. Comparing measurements between upstream measuring point WE02 and upstream one WE10 proves that accumulation of pollutants like TN and TSS which directly affects the eutrophication conditions of the lake takes place in the direction of flow into the lake as shown in figure (5-29a) and (5-29b).

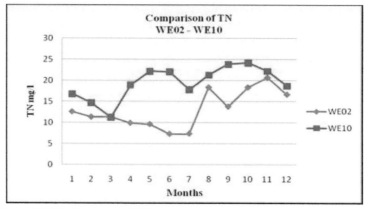

Figure (5-29a): Comparison of TN average conc. measurements between upstream and downstream measuring locations in the catchment

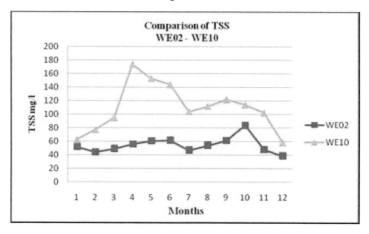

Figure (5-29b): Comparison of TSS average concentration measurements between upstream and downstream measuring locations in the catchment

The interpretation of the catchment condition shows high rates of nutrients leached from the catchment through drainage water into the main drainage system. These nutrients accumulate in the direction of flow until the entrance to the lake. At the downstream zone the nutrients also come from the surrounding fish farms around the lake boundaries. It is expected that the level of nutrients in the lake increases due to all these input sources, and this is shown in the modelling part of the main water quality indicators of chapter 7. It has to be mentioned that it was not possible to find any available historical data sets for Lake Edko to be used as base line for investigation of both hydrodynamics and water quality of the lake. For basic general water quality parameters concentrations in the lake, there were two published papers (Shakweer, 2005) and (Shakweer 2006) through the Oceanography and Fisheries Institute in Alexandria. The papers showed general concentrations of parameters and seasonal variations through site investigations in lake different zones, but also with no historical referenced data. For water quality modelling in shallow lakes it is very important to have historical geographically referenced data sets for the analysis of spatial and temporal variations of parameters concentrations. Therefore it is not only important to monitor the catchment locations but to have fixed monitoring locations for both hydrodynamic and water quality parameters within the lake area.

6. WATER AND NUTRIENTS MASS BALANCE

OF CATCHMENT - LAKE SYSTEM

Water and nutrients balance calculation is an essential step in water quality modelling. This chapter gives a general overview of the water budget of Lake Edko. The mass balance of the nutrients loads entering the lake from the catchment was also calculated and explained in this chapter. Evapotranspitation from the vegetation buffer zone around the lake and the water body are of concern in the calculation of water budget. The chapter focuses on the different types of fertilizers used in the agriculture within the catchment, and the loads of nitrogen and phosphorous discharged into the lake water body.

6.1. INTRODUCTION

A hydrologic or water balance is of considerable importance in water quality analyses and management. A water budget for a lake is an expression of the conservation of water mass for the lake, and can be simply stated as the inflow equals the outflow plus the change in storage of the lake for a given period of time (Lee and others, 1991, p. 14). Construction of a hydrologic budget involves a description and quantification of inputs and outputs. The basic objective of conducting the mass balance is to account for the water gains or losses for the lake, or for some sub-area of interest - and to give a description of the mechanisms by which water is redistributed within the lake. A thorough study of water balance evaluates all the processes which contribute to the inputs or outputs of water to or from the lake system. Calculation of water balance of the lake system is a requirement for the estimation of the lake residence time. The average length of time water remains within the boundaries of an aquatic system is a key parameter controlling the system's biological and chemical behavior. This time scale provides a first order description of the multiple and complex processes that drive the transport of the water and dissolved or suspended substances. Another important item which depends on the water balance of the lake is the nutrients balance. This depends on integrating the estimated inflow and outflow from the water budget with the water quality data.

This chapter gives an overview of the estimated water budget of the lake, the calculation of the residence time and the estimation of the nutrients mass balance, specifically for nitrogen and phosphorous compounds. The calculations are limited to the available historical data and discharge field measurements.

6.2. LAKE WATER BALANCE

The source of a lake's water supply is very important in determining its water quality and in choosing management practices to protect that quality. If precipitation is the major water source, the lake will be acidic, low in nutrients, and susceptible to acid rain. (This includes many seepage lakes.) If groundwater is the major water source, the lake is usually well buffered against acid rain and contains low to moderate amounts of nutrients. (This includes all groundwater drainage lakes and some seepage lakes.) Local septic systems or other groundwater contamination could cause problems. Water exchange is fairly slow.

If streams are the major source of lake water, nutrient levels are often high and water exchange takes place more rapidly. These lakes have the most variable water quality depending on the amount of runoff and human activity in the watershed. Managing the watershed to control nutrients and soil that enter the lake is essential for protecting water quality. Controlling water that runs from the land's surface or streams into the lake is important for drainage lakes and impoundments, and some seepage and groundwater lakes. Protecting groundwater quality is particularly important for seepage and groundwater drainage lakes. Watershed management becomes especially critical in impoundment lakes. Lake managers measure inflow and outflow to determine a lake's water budget. A general and simple hydrologic budget equation for a given water body such as a lake is given by:

$$dV/dt = Q_{in} - Q_{out} + PA_s - E_vA_s$$

where V = lake volume [L^3],

A_s = lake surface area [L^2],

Q_{in} and Q [L^3/T] represent net flows into and out of the lake due to tributary inflows and releases, P [L/T] is the precipitation directly on the lake,

Ev [L/T] is the lake evaporation,

In other words, the rate of change in storage of the volume of water in or on the given area per unit time is equal to the rate of inflow from all sources minus the rate of outflows.

Inputs – outputs = Δ Storage

The input or inflows to a lake may include surface inflow, subsurface inflow, and water imported into the lake. The outputs may include surface and subsurface outputs and water exported (e.g. water supply) from the lake. Some of these flux terms are estimated from the literature (e.g. evapotranspiration) while others are empirical (e.g. discharge at the gauging station). Most components of a water budget are easily measured or estimated. Ground-water flow is the exception, and is an important and largely overlooked component of water budgets because it is the most difficult to quantify as it cannot be measured directly.

In the study catchment of Edko one of the important steps needed as a reference to the mathematical modelling, was to calculate and estimate the water budget of the lake. That was done based on the available collected data from different sources in addition to the two field work survey missions that were done in the study area. Inflow to the lake are measured on monthly basis, but for the outflow from the lake into the sea through the exit channel, no fixed temporal records were present but some field measurements were taken for the purpose of average estimate. Therefore the calculation was done on daily average flows due to the data limitation at the outflow location.

The following terms were evaluated to generate the summary water budget for Lake Edko:

Inputs

1. Surface discharge from two main drains (Edko and Barseek)

Surface-water flow is the largest inflow component of the water budget. Average monthly data exist for the drainage inflow into the lake, but due to the lack of similar measurements at the exit channel, field measurements were taken. The average input discharges measured from the main surface drain Edko during field measurement of 4.01 M m^3/day, in May 2005 were used for calculation. Another measurement of 5.36 M m^3/day was taken in November for validation. Based on hourly discharge rates at Barseek pump station given by DRI, the average discharge rate of Barseek pump station is 0.7 *10^6 m^3/day.

Outputs

1. Outflow from exit channel to the sea (Bougaz Maadia)

To define the interaction between the sea and the lake, the outflow measurements should be measured on a fixed temporal basis, such as monthly, in line with the inflow measurements into the lake. The only available data for the outflow were measurements at the exit channel during field work during May 2005. The average of these measurements was 6.25 M m^3/day. Another field measurement exercise was undertaken in November 2007 in order to validate the calculation; an outflow value of 6.67 M m^3/day was obtained.

2. Evapotranspiration losses

Evapotranspiration is the combined water loss as vapour via evaporation and transpiration from plants. Because the contributing individual processes are difficult to measure separately, they are usually considered together. Evapotranspiration is a complex function of many environmental factors (e.g., windspeed, humidity, temperature, day length, radiation intensity) and has been estimated using various empirical (observation-based) relationships (Thornthwaite 1948, Penman 1948, Thornthwaite

& Mather 1955). For simplified calculations, evapotranspiration is divided into two components: evaporation from open water surface of lake and fisheries and the evapotranspiration from lake vegetation.

a) From water body and fisheries

The lake system has a total approximate area of 124 km^2 and includes the lake water body, the fisheries and the vegetation. These areas are calculated from the developed ARC-GIS digital maps of the study catchment and are divided as follows:

Area of water body = 19 km^2
Area of fisheries = 66.4 km^2
Area of vegetation = 38.7 km^2

The rate of evaporation depends on many factors, including temperature, the amount of solar radiation, vapour pressure, and wind speed. Lake evaporation is often estimated from regional pan-evaporation data by applying a pan coefficient, defined as the ratio of the theoretical free-water surface evaporation to pan evaporation (Farnsworth and others, 1982). Table (6-1) shows the average actual daily and monthly evaporation rates in the Western Delta region

Table (6-1): Average daily evaporation rates in Western Delta region

Mean	Jan	Feb	Mar	Apr	May	Jun	July	Aug	Sep	Oct	Nov	Dec
Daily (mm)	4.1	4.6	5.6	5.6	5.8	5.8	5.6	5.6	5.9	5.8	4.4	3.8
Monthly (mm)	127	129	174	168	180	174	174	174	177	180	132	118
Monthly volumes $*10^6 (m^3)$	2.41	2.45	3.30	3.19	3.42	3.31	3.30	3.30	3.36	3.42	2.51	2.24

b) From lake surrounding vegetation

Evapotranspiration (ET) refers to water lost to the atmosphere before leaving the lake via deep seepage or surface outflow. Much of this is attributable to the water vegetation surrounding the water body. In order to estimate this loss, two pieces of information are necessary: 1) the area of vegetation drawing water more or less directly from the lake, and 2) the rate at which water is drawn by this vegetation. ET from vegetation may be slightly higher than evaporation from water body because: 1) interception of solar radiation exceeds that expected based on the area of vegetation which if not shaded receives extra solar radiation when the sun is low in the sky; enhancing evaporation from leaf surfaces (Hansen 1984). In order to be conservative in the calculation of the water loss from the system, a factor of 1.2 can be used to estimate the evapotranspiration from the lake vegetation based on values published in the literature. The calculations for Lake Edko are shown in Table (6-2).

Based on the simplifying assumptions made in calculating some of the parameters, the following potential sources of the difference in storage (**Δ Storage**) have been identified;

- Groundwater loss and gain to the lake were assumed to be negligible. This should however, be verified with field measurements or through a review of the geology in the area, and referring to other calculations done for similar lakes in the area, since the ground water table in the Delta region is common to them all.
- Inflow and outflow measurements were discrete measured values at specific times during field work; their integration this should be verified by further fixed temporal field measurements.

The estimated water budget shows that the difference between the input and output discharges is $2.22*10^6$ m^3/day. This value is assumed to be mainly subsurface exchange with groundwater.

Since the outflow from the lake to the sea exceeds the inflow rate, this implies an inflow from ground water.

Table (6-2): Water balance estimates for Lake Edko

Item	Discharges	Measured & Calculated values (Million m³/day) 2005	Measured & Calculated values (Million m³/day) 2007
	Inputs		
1	Inflow from Edko Drain	4.01	5.36
2	Inflow from Barseek Drain	0.7	0.7
	Total	**4.71**	**6.06**
	Outputs		
1	Outflow from sea exit	6.25	6.67
2	Evaporation losses from water body	0.11	0.084
3	Evaporation losses from fisheries	0.39	0.29
4	Evapotranspiration from vegetation	0.27	0.2
	Total	**7.02**	**7.2**
	Δ Storage	**2.31**	**1.18**

Groundwater is found at shallow depths. Before the implementation of tile drainage systems, it fluctuated between 0 and 1 m below the ground surface (RIGW, 1986). Since EPADP installed tile drainage systems throughout most of the area the groundwater table has been lowered and fluctuates between 1 and 1.6 m below the surface. In most areas, freshwater layer overlays saline water, towards the north, the freshwater lens becomes thinner and more saline. Waterlogging occurs locally, notably along the continuous flowing canal sections (seepage and leakage).

This groundwater inflow/outflow should be calculated based on data for the groundwater characteristics in the Nile Delta complemented by modelling. A detailed water budget study was conducted for the coastal Lake Burullus in the middle Delta region. This showed groundwater inflow to the lake (Shinnawy, 2000). For the estimation of the water balance for Lake Edko, the results could be accepted within the described data limitations. We conclude that the calculation of groundwater recharge to the lake system and the evapotranspiration are two main issues for water balance calculation in shallow lakes.

6.3. CALCULATION OF LAKE RESIDENCE TIME

The average length of time water remains within the boundaries of an aquatic system (hydraulic residence time) has been proposed in the literature as an important parameter with which to explain a range of water quality phenomena such as the variability in lake eutrophication processes, thermal stratification, isotopic composition, alkalinity, dissolved organic carbon concentration, elemental ratios of heavy metals and nutrients, mineralization rates of organic matter, and primary production (see Monsen et al. 2002, for a list of references). The basic concepts on transport time scales and their application to coastal environments are laid out in the works of (Zimmerman 1976) and (Dronkers and Zimmerman 1982). There, the most commonly used terms to measure the retention of water or scalar quantities transported in the water are carefully defined, and suitable experiments are presented to calculate the defined time scales, with examples for coastal environments. The experiments go further, suggesting that the hydraulic residence time is a key parameter controlling the structure of aquatic ecosystems and the extent to which these systems are self-organized or dominated by outside influences. The lake's size, water source, and watershed size are the primary parameters that determine the retention time. Rapid water exchange rates allow nutrients to be flushed out of the lake quickly. Such lakes respond best to management practices that decrease nutrient input. Impoundments, small drainage lakes, and lakes with large volumes of groundwater inflow and stream outlets (groundwater drainage lakes) fit this category. Longer retention times occur in seepage lakes with no surface outlets. Average retention times range from several days for some small impoundments to many years for large seepage lakes.

Nutrients that accumulate over a number of years in lakes with long retention times can be recycled annually with Spring and Fall mixing. Nutrients stored in lake sediments can continue to be released, even after the source of nutrients in the watershed has been controlled. Thus, the effects of watershed protection may not be apparent for a number of years. Nevertheless, lakes with long retention times tend to have the best water quality as shown by the lower levels of the plant nutrient phosphorus in Table (6-3). Better water quality is a consequence of the greater depths of the lakes and their relatively smaller watersheds.

Table (6-3): Several characteristics of lakes with different retention times

Retention Time in days	0-14	15-60	61-180	181-365	366-730	>730
Mean depth (ft)	6	8	11	11	13	23
Maximum depth (ft)	16	21	25	27	35	57
Mean total phosphorous (µg/l)*	94	85	56	48	33	25
Mean DB:LA ratio**	1166	142	42	15	8	6

(*Adapted from Lillie and Mason, 1983.*)
*Summer values; µg/l = micrograms per liter or parts per billion
**DB:LA = Drainage basin/lake area

Drainage basin/lake area ratio (DB:LA)

The size of the watershed (drainage basin) feeding a lake relative to the lake's size (area) is an important factor in determining the amount of nutrients in a lake. Table (6-3) shows this relationship for a sample of Wisconsin lakes, (USA). Lakes with relatively large drainage basins usually have significant surface water inflow. This inflow carries more nutrients and sediments into these drainage lakes or impoundments. By definition, seepage lakes have small drainage basins, more groundwater flow, and fewer nutrients from runoff. Groundwater drainage lakes typically have an intermediate-sized drainage basin. Low-ratio lakes (small drainage basin and large lake area) have high retention times while high-ratio lakes have short retention times. Drainage basin:lake area ratios can be used to estimate a lake's retention time. Table (6-4) shows the geomorphologic characteristics of Lake Edko and the lake residence time. Based on the average annual inflow rate of 50 m^3/sec, the residence time for Lake Edko is estimated to be 5.3 days.

Table (6-4): Geomorphologic Characteristics of Lake Edko,

Water surface area (km2)	Catchment area (km2)	Coast length (km)	Average Length of lake (km)	Mean width (km)	Mean depth (m)	Maximum depth (m)	Lake volume (m3)	Annual inflow (m3/day)	Residence time (d)
19	2048	33	8	3	1.2	2.5	228*10^5	4320000	5.3

6.4. ESTIMATION OF NUTRIENTS LOADS

Nutrients are fundamental for plant growth in all ecosystems, as much in water as on land. The natural nutrient status of a water body, such as a lake, depends on its size, depth, geology, retention time and the use of land in its catchment area. Some soils release more salts and nutrients than others and therefore different water bodies in their natural state can be at different trophic, i.e. nutrient concentration, levels. As mentioned above, an increase in available nutrients or the nutrient enrichment of a water body is referred to as a eutrophication condition. Nitrogen, phosphorus, carbon and, in some cases, silicon are the nutrients of most concern in relation to eutrophication. However, because phosphorus is the most limiting nutrient in the fresh water environment, it attracts the most attention in this case.

6.4.1. Sources of nutrients in Lake Edko

Lakes, and shallow lakes in particular, are aquatic ecosystems often with an ecological structure which consists of a great diversity of plant and animal life. This structure is finely balanced and the species are inter-dependent. Birds and mammals rely for their food supply not only on the total productivity of the species living in the lake but on the presence of a particular species, which in turn is subject to the availability of nutrients. The microscopic algal communities, phytoplankton, are the first link in the food chain for aquatic animals and fish. These and other floating plants acquire their nutrients directly from the water, whereas the macrophytes that root into the lake bottom can obtain nutrients from the sediments. These larger plants are crucial for the maintenance of the oxygen level in the water.

In the case of Edko Lake the situation is considered complicated because the lake receives drainage water rich in fertilizers from the catchment in addition to unrecorded discharges from fish ponds. Therefore, Agriculture is considered the largest source of nutrients into the lake water. This is due to the excessive use of fertilizers rich in phosphorous and nitrogen compounds. Before we discuss the analysis of the nutrients balance in Lake Edko it is important to give a summarized overview of the fertilizers application and types used in Egypt.

6.4.2. Fertilizers Types Used in Egypt

Agricultural land accounts for only 3.5 percent of the land area of Egypt. Two thirds of the agricultural land is alluvial soil, fertilized for thousands of years by the Nile floods, and one third is land recovered since the 1950s. Rainfall is minimal and almost all the agricultural land is irrigated. Soil salinity and water logging are important problems in the reclaimed areas. Sprinkler irrigation and drip irrigation are commonly used on the recovered area, and fertilizers are used on 13 percent of the land. There are up to three harvests per year, the overall cropping intensity being 180 percent. Crop yields and rates of fertilizer use are therefore relatively high. The construction of the High Aswan Dam reduced the quantity of suspended materials deposited on the soil during floods, which used to restore the fertility of Egyptian soils for thousands of years.

Egypt has a long tradition of using mineral fertilizers; its first use of Chilean nitrates dated back to 1902. For over thirty years, all mineral fertilizers were imported, until the local production of phosphate fertilizers started in 1936. The production of nitrogen fertilizers began in 1951. No potash fertilizers are produced in Egypt due to the lack of resources, although it was reported recently that some local potash deposits had been found (FAO, 2005). The demand for food and other agricultural commodities is increasing in Egypt due to the increase in the population and improvements in living standards. Efforts continue to improve crop productivity and quality. The breeding of new high yielding varieties and the development of better agricultural practices are some of the measures aimed at increasing agricultural production to meet the increase in demand.

Appropriate fertilization is one of the most important agricultural practices for achieving the objectives on production. Evaluation of the best source of nutrients, optimum rates of fertilization, suitable timing and proper fertilizer placement are necessary for efficient fertilizer management. In Egypt, there are several traditional practices that are commonly implemented and which play a major role in restoring and maintaining soil fertility. Among these practices are:

- Planting berseem clover as a winter fodder crop before the cotton crop, which provides a green manure that is ploughed in after taking one or two cuts.

- Incorporating farm yard manure (FYM) into the soil during seedbed preparation. This is usually done before planting an important cash crop such as cotton.
- Including a legume crop in the crop rotation such as faba bean, clover and soybean, which have a positive effect on soil fertility and contribute to part of the nitrogen requirement.

Improvements in product quality and production efficiency, either already achieved or planned, permit the domestic fertilizer industry to compete successfully with most imports. Figure (6-1) provides details of the production, import, export and consumption of fertilizers in the period between 1998 and 2002. Urea is produced domestically and part is exported. For example, in 2002 the total domestic production of urea was five million tonnes of which 23 percent was for export and 77 percent for the domestic market.

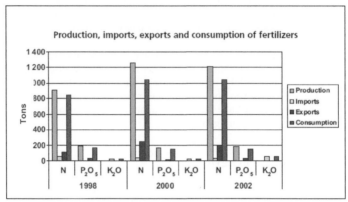

Figure (6-1): production, imports, exports and consumption of fertilizers
in the period between 1998 and 2002 (Source: MALR, 2003)

The main types of fertilizers used in Egypt are:
Nitrogen
- urea (46.5 percent N)
- ammonium nitrate (33.5 percent N)
- ammonium sulphate (20.6 percent N
- calcium nitrate (15.5 percent N)

Phosphate
- single superphosphate (15 percent P2O5)
- concentrated superphosphate (37 percent P2O5)

Potassium
- potassium sulphate (48 to 50 percent K2O)
- potassium CHL-aoride (50 to 60 percent K2O)

Mixed and compound fertilizers containing N, P, K, Fe, Mn, Zn and/or
- Cu in different formulations for either soil or foliar application. The
- Micronutrient may be in either mineral or chelate form.

In the agricultural year 2002/03, the crop areas in the old and new lands in Egypt amounted to 5.1 and 1.4 million ha respectively. According to the "indicative cropping pattern", the area allocated to various crops and the recommended rates of nitrogen and phosphate fertilizers, it is estimated that the country needs 1.1 million tonnes of N and 364 thousand tones of P2O5. The status of potash and the micronutrients in most Egyptian soils is different from that of N and P. As a consequence, it has been decided to determine requirements for these nutrients by taking the consumption of the previous year and increasing it by about 10 percent. Figures (6-2) shows the estimates of the areas and fertilizer requirements of "new" and "old" land respectively up to 2017.

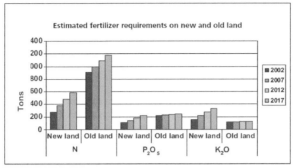

Figure (6-2): Estimates of the areas and fertilizer requirements of "new" and "old" land up to 2017 *(Source: MALR, 2003)*

Efforts are being made to increase the composting of agricultural residues as a source of plant nutrients, in order to contribute to the improvement of the physical properties of the soil and the protection of the environment. According to the FAO report, 2005, mineral fertilizers, especially nitrogen, phosphate and potash are being applied to an increasing extent in Egypt. Figure (6-3) shows that the consumption of nitrogen and phosphate fertilizers has tripled during the last 30 years.

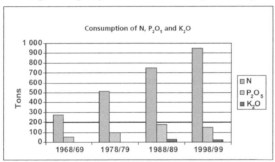

Figure (6-3): Consumption of nitrogen and phosphate fertilizers during the last 30 years *(Source: Taha, 2000)*

The input of nutrients from different sources, such as soil, air, water and other sources, including organic manures, should be equal to the amount removed or taken up by the crop. If the quantities from these sources are not sufficient for the crop to reach the target yield, the difference should be added as fertilizer. Insufficient amounts of nutrients result in a, loss of yield. Excessive amounts represent a waste of resources, possibly a decrease of yield and could be an environmental hazard, as identified in the case of nitrogen (Fawzi, 1992).

From the information given above on the use of fertilizers in Egypt, it is obvious that a huge amount of fertilizer is drained to the surface waters as a residual from its application. Most studies carried out to evaluate different sources of phosphate have focused on the direct effect of these sources. Very little work has been done on evaluating the residual effect of these sources by means of long-term experiments. The fertilizing value of various sources of phosphate should include both the direct and the residual effects of the different sources under different cropping systems. Because nitrogen fertilizers are the most used fertilizers in Egypt, studies should focus on the rate of nitrogen fertilizers applied to the main crops, using different sources of nitrogen, time and method of application, in order to increase the efficiency of the fertilizers and to reduce nitrogen losses (Hamissa, 2000) to surface waters.

6.4.3. Effect of Nutrients on Edko Drainage Catchment and Lake System

The Lake Edko drainage catchment receives water from several sub-catchments through the six pump stations as mentioned above. The main nutrients entering the lake are phosphorous and nitrogen compounds.

Phosphorus promotes excessive aquatic plant growth. Phosphorus originates from a variety of sources, many of which are related to human activities. Major sources include human and animal wastes, soil erosion, detergents, septic systems and runoff from farmland or lawns. Phosphorus provokes complex reactions in lakes. An analysis of phosphorus often includes both *soluble reactive phosphorus* and *total phosphorus*. Soluble reactive phosphorus dissolves in the water and readily aids plant growth. Its concentration varies widely in most lakes over short periods of time as plants take it up and release it. Total phosphorus is considered a better indicator of a lake's nutrient status because its levels remain more stable than soluble reactive phosphorus. Total phosphorus includes soluble phosphorus and the phosphorus in plant and animal fragments suspended in lake water. Ideally, soluble reactive phosphorus concentrations should be 10 µg/l (micrograms per liter) or less at Spring turnover to prevent summer algae blooms. A concentration of 10 micrograms per liter is equal to 10 parts per billion (ppb) or 0.01 milligrams per liter (mg/l). A concentration of total phosphorus below 20 µg/l for lakes and 30 µg/l for impoundments should be maintained to prevent nuisance algal blooms.

In most cases, the amount of nitrogen in lake water corresponds to local land use. Nitrogen may come from the application of fertilizer and animal wastes on agricultural lands, human waste from sewage treatment plants or septic systems, and lawn fertilizers used on lakeshore property. Nitrogen may enter a lake from surface runoff or groundwater sources. Nitrogen exists in lakes in several forms. Analyses usually include nitrate (NO_3) plus nitrite (NO_2), ammonium (NH_4), and organic plus ammonium (Kjeldahl nitrogen). Total nitrogen is calculated by adding nitrate and nitrite to Kjeldahl nitrogen. Organic nitrogen is often referred to as biomass nitrogen. Nitrogen does not occur naturally in soil minerals, but is a major component of all organic (plant and animal) matter. Decomposing organic matter releases ammonia, which is converted to nitrate if oxygen is present. This conversion occurs more rapidly at higher water temperatures. All inorganic forms of nitrogen (NO_3, NO_2 and NH_4) can be used by aquatic plants and algae. If these inorganic forms of nitrogen exceed 0.3 mg/l (as N) in spring, there is sufficient nitrogen to support summer algae blooms.

6.4.4. Nitrogen Compounds in Lake Edko

Based on earlier research done on Lake Edko, (Shakweer, 2006) indicated that the highest concentrations of ammonia were found at either the eastern or western areas of the lake during winter compared with other seasons of the year. These concentrations decreased gradually to reach the lowest values during summer. These decreased concentrations may be due to the consumption and assimilation of ammonia by aquatic plants. Hutchinson (1957) found that the sudden fall of ammonia is accompanied by a great increase in aquatic plants. In Lake Edko the maximum growth of aquatic vegetation is in summer. (Samman, 1974). On the other hand the highest concentrations of this inorganic compound in the drainage water were recorded during winter. Kenawy (1974) in her study on the chemistry of lake Edko water indicated that the concentrations of ammonia ranged from 0.12 to 0.19 µg/L with an average of 0.15 µg/L The average concentrations of both nitrites and nitrates tend to fluctuate with the same pattern as ammonia for which the highest concentrations were found in winter. The most important factors affecting the concentration and distribution of nitrate in the lake are the discharge of drainage water, dissolved oxygen, decomposition of organic remains regeneration from the bottom sediments and assimilation by aquatic plants.

This means that the main source of ammonia to the lake is the drainage water outflow through Edko and Bersik drains. The increased concentrations of ammonia in the drainage water during winter can be attributed to the application of large amounts of ammonium nitrate as fertilizer during this period of the year. The type and rate of application of the fertilizer depends to a great extent on the variations in agricultural practices from one season to another. However it is obvious that the average concentration of ammonia in the lake water has obviously increased during the last 20 years.

Higher concentrations of ammonia, nitrite and nitrate existed at the eastern area of the lake. This area can be attributed to the increased concentrations of these nitrogenous compounds in Edko drain water which outflows at the east of the lake where the average concentration of ammonia was 24.23 µg/L and the average concentration of nitrite was 9.36µg/L. It seems on the other hand that ammonia and nitrites were oxidized with higher rates in the western area of the lake in comparison with those in the eastern area. Therefore higher concentrations of nitrates existed in this part of the lake.

It can be observed from the data that the dominant compound throughout the whole year was nitrate. The dominance of nitrate in both the lake and drainage water indicates that the dissolved oxygen in the water was sufficient to oxidize most of the ammonia and nitrites. From the seasonal variations of these percentages it can be observed that ammonia dominated the nitrogenous compounds in the eastern area during winter and spring. On the other hand, it was dominant during spring only in the western area, where the nitrate comprised 83.11%, 63.87%, and 61.60% during summer, autumn and winter respectively (Shakweer, 2006).

It can also be observed that ammonia dominated the nitrogen compound in the drainage water in Edko drain during spring and autumn, comprising 48.53% and 55.75% of the total nitrogen compounds during these two seasons. In Barseek drain nitrate was the most dominant, where it constituted 48.03%, 71.77% and 68.12% of the total nitrogen compounds during spring, summer and winter in respective. This indicates that the water from Barseek drain provides the lake with more nitrogenous compounds than Edko drain. It should be pointed out that most of the domestic and fish farm wastes in the southern part of Lake Edko are discharged to the lake through Barseek drain while the drainage water of Edko drain is mainly composed only of agricultural wastes. It was expected that due to the difference in the discharge rates between the two drains ther would be higher concentration values of nutrients from Edko drain because it drains a cultivated area of about 240 feddans (about1000 km²). But the data shows thatconcentrations from Barseek drain are higher. This can be attributed to the high nutrient concentrations from the fish farms.

It must be pointed out also to the fact that the distribution of ammonia in the lake water is affected by several factors such as temperature, dissolved oxygen, decomposition of organic matter and assimilation by plant organisms, in addition to wind as another important hydrodynamic driving force, so the nutrient processes in the lake are complex in nature.

6.4.5. Phosphorous Compounds in Lake Edko

The phosphorus cycle in lakes has been recognised by Harvey (1974) as being very different way from that in the sea. He points out that in lakes algae fall to the bed, due to the zooplankton population being too small to eat, crush or digest them while in suspension. However the algae at the bed are very slowly decomposed by bacteria and protozoa. Therefore large quantities of phosphorus remain locked up in the algal deposit. The water of lakes is in consequence often rich in salts containing nitrogen but very poor in phosphate. However phosphorus is considered as one of the important nutrient element in the aquatic habitate. This element is a one of the limiting factors controlling the growth and reproduction of phytoplankton. The environmental significance of phosphorus arises out of its role as a major nutrient for both plants and micro organisms (Vanloon and Duffy, 2000).

Studies show that higher concentrations of phosphate were recorded in Edko drain than inBerseek. The yearly average concentration of such a phosphorus compound was found to be 12.11 and 10.87 µg/L in the two drains respectively. On the other hand the average concentrations of phosphate in the water of the eastern and western areas of the lake were found to be 8.08 and 6.59 µg/L respectively. It is believed therefore that the higher concentration of phosphate in the eastern area is due to the Edko drain. Similarly, the decreased concentrations in the southern area can be attributed to the lower concentrations of this nutrient salt in Barseek drain water.

The average concentrations of phosphate in the drains were mostly higher than the concentrations in the lake. This means that the drains are the main suppliers of phosphorus compounds to the lake. Runoff from the cultivated land near to the lake provides a major contribution of nutrient salts to the lake. The smallest concentrations of phosphates in the lake were recorded in summer. This decrease of phosphate concentration in the lake can be attributed to the maximum growth of phytoplankton and rooted plants in the lake during summer and consequently the uptake of large amounts of such nutrient salt.

Therefore it is concluded from earlier studies that the drainage water is generally rich in dissolved nutrient salts with special reference to the periods when fertilizer consisting of nitrogen and phosphorus salts is applied over the cultivated areas around the lake. The highest concentrations of ammonia were found in the lake during winter. Decreased concentrations were detected during spring corresponding to increased aquatic vegetation. Higher concentrations of nutrient salt exist in the eastern part of the lake where Edko drain discharges to the lake. In the next section, the analysis of the average nutrient concentrations flowing into the lake from the drains is shown for the year 2005-2006, and the estimated loads of nutrients into the lake are calculated based on these values. The concentrations of total nitrogen and total phosphorous compounds are given in Table (6-5).

Table (6-5): Average concentrations and loads of nutrients from Edko drain to the Lake

Month	TN (mg/l)	TP (mg/l)	Q (m³/sec)	Loads (TP) Ton/Month	Loads(TN) Ton/month
Jan	40.2	1.5	38.63	150.18	4024.82
Feb	32.524	0.607	28.47	44.79	2400.03
Mar	35.67	0.896	46.20	107.29	4271.24
Apr	26.73	0.48	50.84	63.25	3522.20
May	34.85	0.69	63.49	113.22	5735.31
Jun	19.03	0.525	60.23	81.96	2970.81
Jul	3.94	0.631	67.09	109.73	685.19
Aug	19.55	0.25	68.30	43.55	3460.76
Sep	15.65	0.69	73.04	130.45	2963.04
Oct	21.89	0.72	43.83	81.79	2486.58
Nov	3.30	0.67	40.36	70.10	345.26
Dec	17.33	0.69	46.35	83.01	2081.84
Total	270.66	8.34	626.82	1079.32	34947.10

7. DEVELOPMENT OF HYDRODYNAMIC, WATER QUALITY AND EUTROPHICATION MODELS

Water quality models are considered key elements in understanding water quality problems for management and decision support processes. Modelling of surface water quality for management purposes at a watershed scale involves different water bodies, such as rivers, drainage networks and lakes. The understanding of water quality problems in such an interacting system needs a proper description of its hydrodynamic status. As a component of the water quality management information system for the Edko drainage catchment and shallow lake, a 1D-2D hydrodynamic model is developed to understand the basic hydrodynamics of the catchment–lake system. This model is coupled with a water quality model of the catchment drains to simulate the temporal and spatial variations of water quality with respect to nutrients, dissolved oxygen, suspended sediments and microbiological parameters. For further analysis of the water quality and eutrophication condition of the lake, a 2D hydrodynamic-water quality model and eutrophication model is developed.

7.1. INTRODUCTION

As explained in the previous chapters, agricultural irrigated watersheds have a complex physical nature in that they include interacting irrigation and drainage networks which may be connected to lakes or lagoons. Therefore, studying surface water quality problems in such watersheds, calls for an integrated modeling approach that takes into consideration the different components of the watershed in both problem assessment and the development of solutions. The developed models are aimed at a better understanding of the hydrodynamics of a watershed where the drainage catchment is connected to a shallow coastal lake, which is considered the main outlet of the drainage water to the sea. A 1D-2D model of the drainage network and the shallow lake is developed using the Sobek 1D-2D mathematical modeling package. This model reveals the basic hydrodynamic properties of the drainage network and its link with the lake water body. For a more detailed description of the hydrodynamics and water quality, a detailed 2D hydrodynamic model is developed for the lake area using the Delft 3D modeling tool. The detailed 2D model of the lake shows the effect of the tidal forces and wind on the lake hydrodynamics and the corresponding effect on the spatial distribution of pollutants in the lake. A 2D water quality model is coupled with the 2D hydrodynamic model of the lake to study the effect of different chemical and biological pollutants discharged into the lake, in addition to the effect of nutrients on the lake eutrophication.

7.2. MODELS BUILDING

The overall hydrodynamic modelling approach used in this research is based on the Sobek and Delft-3D modeling software developed by WL|Delft Hydraulics. Sobek is an integrated package for modeling river, urban or rural water networks, and the Sobek 1D channel flow module can be integrated with the 2D Delft-FLS model. It is therefore possible to implement a 1D-2D implicitly coupled modeling system. This provides a high-quality tool for modeling irrigation systems, drainage networks, natural streams and rivers in lowland and hilly areas. The functionality contained in the hydrodynamics of two-dimensional flow is part of the Sobek-Rural model package. This functionality includes the one dimensional river/channel flow and two-dimensional flow and is called "**Sobek 1D-2D**". The numerical aspects of 1D modeling include the water flow and pollutants transported by the flow. The water flow is computed by solving the complete Saint Venant equations. The mass conservation equation is applied at the calculation and connection points (1D nodes), and the momentum equation is applied at the reaches between two calculation nodes. The Saint Venant equations (7-1), (7-2) expressed in their non-conservative form per unit width of channel are as follows:

Continuity equation

$$\frac{\partial \zeta}{\partial t} + \frac{\partial (uh)}{\partial x} = 0 \tag{7-1}$$

Momentum equation

$$\frac{\partial u}{\partial t} + u\frac{\partial u}{\partial x} + g\frac{\partial \zeta}{\partial x} + cf\frac{u|u|}{h} = 0 \tag{7-2}$$

where
ζ: Water level (based on reference level) defined as $\zeta = h + zb$ [m];
h: Local water depth [m];
zb: Local bottom level [m]
u: Flow velocity [m/s]
cf: Friction coefficient [-]

For two dimensional flow, two momentum equations are calculated in the x- and y-directions, together with the continuity equation 2D. The momentum equations (7-3) read:

$$\frac{\partial u}{\partial t} + u\frac{\partial u}{\partial x} + v\frac{\partial u}{\partial y} + g\frac{\partial \zeta}{\partial x} + g\frac{u|v|}{C^2 h} + au|u| = 0$$

$$\frac{\partial v}{\partial t} + u\frac{\partial v}{\partial x} + v\frac{\partial v}{\partial y} + g\frac{\partial \zeta}{\partial y} + g\frac{v|v|}{C^2 h} + av|v| = 0$$

(7-3)

where:

u : Velocity in x-direction [m/s]
v : Velocity in y-direction [m/s]
V : Velocity: [m/s]
ζ : Water level above reference [m]
C : Chézy coefficient [$m^{1/2}$/s]
d : Depth below reference [m]
h : Total water depth $h = \_ + d$ [m]
a : Wall friction coefficient [1/m]

The Delft3D modeling system developed by WL|Delft Hydraulics consists of different sub-models or modules that are capable of simulating time and space variations due to different phenomena. The Delft3D-FLOW hydrodynamic modeling module is a multi-dimensional (2D or 3D) hydrodynamic and transport program that calculates non-steady flow and transport phenomena resulting from tidal and meteorological forcing (WL|Delft Hydraulics, Delft3D-FLOW User Manual 2005). This module is used for detailed modeling of Lake Edko hydrodynamics. The system of equations in Delft3D-FLOW consist of the shallow water equations derived from the 3D Navier-Stokes equations for an incompressible fluid using the shallow water and Boussinesq assumptions as well as the continuity equation.

The Delft3D-WAQ module is used for water quality modelling coupled with the hydrodynamic module. Delft3D-WAQ is a 3-dimensional water quality modelling framework. It solves the advection-diffusion-reaction equation on a predefined computational grid for a wide range of model determinands. Delft3D-WAQ allows great flexibility in the substances to be modeled, as well as in the processes to be considered. The Delft3D-WAQ is not a hydrodynamic model, so information on flow fields has to be provided by the hydrodynamic module. Coupling of the hydrodynamic model and the water quality module is done to develop the water quality model.

7.3. BASE MODEL OF THE CATCHMENT SHALLOW LAKE SYSTEM: LINKING OF 1D-2D HYDRODYNAMIC MODELLING

The hydrodynamic model developed for the drainage network-lake system is shown in Figure (7-1). It is composed of a 1D network representing the two main drains entering lake Edko, and a 2D grid component representing the lake area. The 1D network is composed of two sub-networks representing the Edko drain and Barseek drains. The main drain has 142 cross sections. The upstream boundary condition consists of a prescribed time series of discharges, and there are six lateral inflows discharging into the Edko main channel. Knowing the measured discharges and water levels, the contributions of the lateral flows from various drain sources are calculated. The lake is considered as the outlet of the drainage catchment. The flows in the lake are simulated on a 2D grid, which consists of 97 grid points in the (west-east) x-direction and (south-north) 53 grid points in the y-direction. The grid size is 100 x 100 m. There are a series of stations for measuring water levels and discharges in the lake. Water level variations and flows in response to wind forcing and tidal effects at the lake surface are simulated along with the inflows and outflows. Water quality parameters of the lake and the assessment of the trophic conditions are also simulated in this model. The shallow water equations are based on the assumption that the depth of flow is small compared with the horizontal length scales involved, resulting in two-dimensional depth-averaged equations. Due to the shallow characteristics of Lake Edko, this assumption is reasonable.

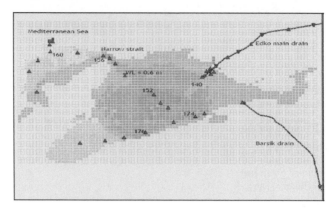

Figure (7-1): Schematisation and setup of 1D-2D model of the lake catchment system

The hydrodynamics of the study area are described by the one-dimensional unsteady flow simulations for the Edko Drain, which has a continuous drainage flow and the two-dimensional description of the simulated water levels, wind driven currents, and tidal flow through the inlet from the Mediterranean Sea. The boundary conditions for the flow model include the time series inflow to the lake from the Edko main drain with its contributions from six lateral tributaries, the inflow time series from the Barseek drain from the south, and the tidal inflow defined as a water level time series at the exit channel to the Mediterranean Sea. The simulation period for the model is selected to be one year to cover the different seasonal variations in the discharges and pollution loads. The simulation is done for two different cases: the first including the inflows from the drains and the tides without the wind and the second including the wind. The simulation results for the Edko main drain under normal conditions reveal fluctuating water levels with an average value of 2.60 m during a period of 2 months as shown in Figure (7-2).

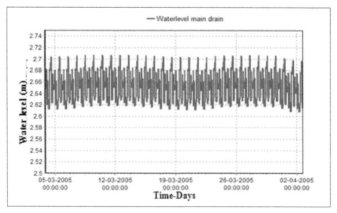

Figure (7-2): Water level variations in Edko main drain

Given the characteristics of the flow regime to the lake from the drains and assuming the influence of no wind for the given period, the results clearly suggest that water level oscillations in the lake as recorded by the stations 140 (DS Edko drain) and 152 (middle of lake) shown in Figure (7-3), are not due to the inflows, but to the tidal wave that propagates from the ocean inlet through the lake.

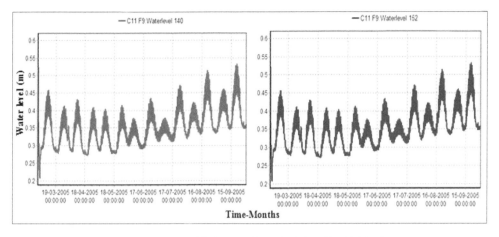

Figure (7-3): Water level oscillations in Edko Lake induced by tidal forcing at two Stations (Sta. No. 140, 152).

A comparison between the water level oscillations at the sea boundary and at the stations in the lake is shown in Figure (7-4) (i.e., Sta. No. 140, 159). This verifies that the water levels at the downstream part of the exit channel to the sea are mainly affected by the tidal level variations.

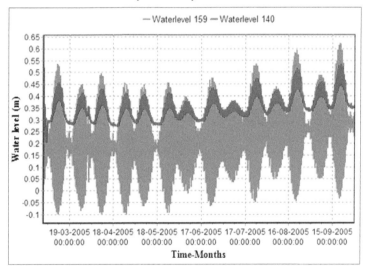

Figure (7-4): Comparison between the water level oscillations at Sea Boundary and in the Lake (Sta. No. 140).

The 1D-2D model revealed the dynamics of the drainage water discharges from the main drains, and the coupling of the sea water intrusion into the lake and water level variations. However, model could not provide a clear definition of the direction of the flow velocities.

7.4. DETAILED 2D HYDRODYNAMIC SHALLOW LAKE MODEL

The detailed 2D hydrodynamic model of the shallow lake was developed using the Delft3D Mathematical model package. The model was developed for a further and more detailed study of the lake hydrodynamics, taking into account the effects of the main flow driving forces which are the wind and tidal forces. The model bathymetry map was developed from field hydrographic survey data processed with Arc-GIS, and then this map was imported into the Delft3D grid generation module RGFGRID to develop a curvilinear calculation grid for the model. Figure (7-5) shows the calculation grid and the depth map.

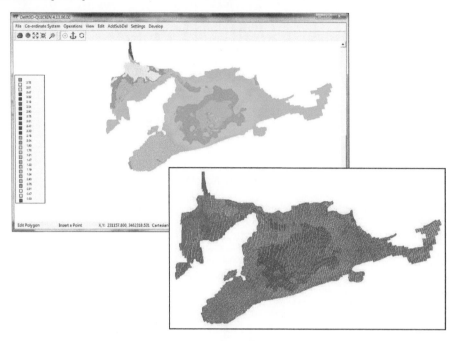

Figure (7-5): 2D model shematisation and bathemetry map developed with Delft3D RGFGRID module

The modelling period was selected to be six months starting from January 2006 till July 2006; this period was based on the time for field work and on the data collected for hydrodynamic modelling and from remote sensing.

Model Boundary Conditions and Physical parameters

In the model area there are three open boundaries. The first boundary is the exit channel to the sea, which connects the lake with the open sea. The tidal water level variation for the simulation was specified at that boundary. The second open boundary is the Edko main drain inlet to the lake, and the third open boundary is the Barseek main drain inlet to the lake.

The inflow discharges from the main drains Edko and Barseek are shown in Figures (7-6 a), (7-6 b). So the model boundary conditions are chosen to be the discharge inflow from the two main drains, and the tidal variation of water level at the sea inlet to the lake. A spatially varied bottom roughness grid was developed using Manning roughness coefficients that range between 0.031 and 0.07.

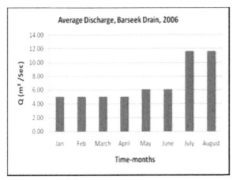
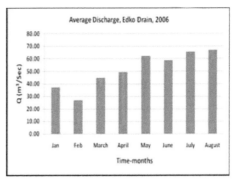

Figure (7-6a, b): Monthly discharges of Edko and Barseek Drains (year 2006).

The tide in the ocean is semidiurnal with amplitude ranging between 0.2 and 1.2 m. Due to a lack of measured data, tidal varying water levels at the exit channel to the sea were developed using theWX-TIDE prediction computer program. Figure (7-7) shows the tidal water level variations for the selected modelling period.

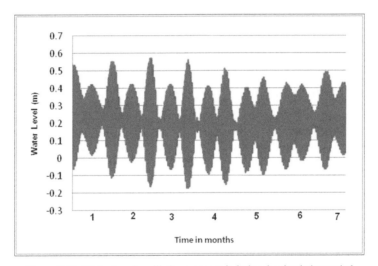

Figure (7-7): Water Level oscillations at sea exit during the simulation period
(January-July 2006).

Wind data used in the model input is based on an hourly time series data collected from the meteorological station at Alexandria port, which is the nearest point to the study catchment at location (31.2 latitude and 29.95 longitude). An analysis of the wind time series for the year 2006 gives the wind rose diagram, which shows the distribution of the wind speed and the frequency of the varying wind directions. It is concluded from this analysis that the prevailing wind direction in this coastal region is NNW with an average speed of 5 m/sec as shown in Figure (7-8).

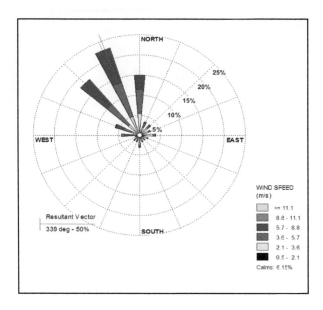

Figure (7-8): wind rose diagram showing the distribution of the wind speed and the frequency of wind direction.

Evapotranspiration is an important physical parameter in the hydrodynamic modeling of lakes. This was taken into consideration in the model development through the application of an absolute flux and net solar radiation heat flux model. This model takes into account the area of the water body, the sky cloudiness and the monthly average data of relative humidity, air temperature and net radiation. The evapotranspiration from water vegetation is accounted for as a percentage of the overall inflow into the lake based on the water balance calculations for the lake. Figure (7-9a, 7-9b, 7-9c)show respectively the monthly average air temperature, relative humidity and solar radiation data for the simulation period.

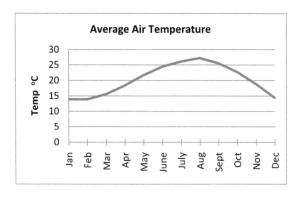

Figure (7-9 a) : Monthly average air temperature during simulation period

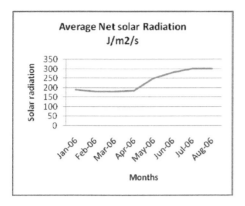

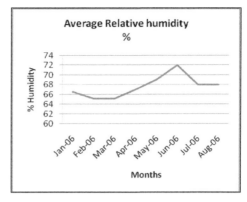

Figure (7-9 b): Average relative humidity during simulation period

Figure (7-9 c): Average solar radiation during simulation period

Various model runs were performed assuming no forcing parameters of wind and tidal flow; other runs were conducted including the discharges from the main drains and a combination of the forcing parameters with the main purpose of understanding the hydrodynamics of the lake and its response to the different forces.

Case I: Effect of Discharges and Tidal Forces

In the first model run, the simulation only takes into account the tidal inflow from the sea and the discharges from the main drains with no wind forcing. The initial water level in the lake is set to 0.4 m. Water level and velocity results are derived at 11 history stations and discharges at three cross sections. These are displayed in Figure (7-10a) and (7-10b). Note that the stations are selected as locations of particular interest to represent the different hydrodynamics parameters in the lake.

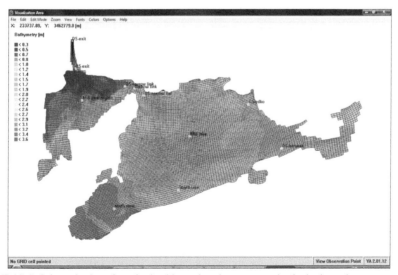

Figure (7-10 a): Schematisation of monitoring history locations representing hydrodynamic parameters of the model

139

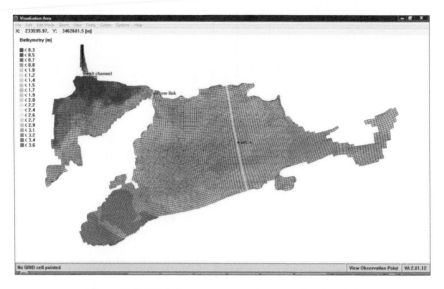

Figure (7-10 b): Discharge measurement cross sections locations

The results from the first run show that the water levels at the mid lake monitoring station range from 0.17 m to 0.57 m with a total variation of 0.4 m, during the simulation period of six months. Figure (6-11) shows the water level oscillations at this location. The effect of the tide on the water levels oscillations near the exit channel area is obvious, as shown in Figure (7-11).

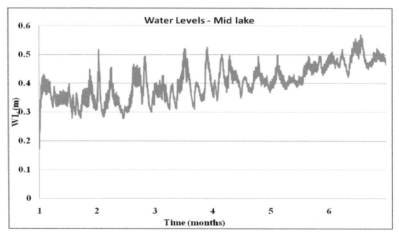

Figure (7-11): Water Levels calculated for the period of six months (Jan-July, 2006) at monitoring location mid lake

From the calculated water levels at different monitoring locations it is also obvious that the water level variation does not differ significantly at locations within the mid area of the lake. However there are distinct variations between the locations around the lake exit where the flow is affected by the tidal forces and around the drain inlets to the lake where the effect of the tidal forces is dominated by the drain discharges. An average total variation of ± 26 cm in water level occurs near the exit channel entrance and this variation decreases as we go into the middle area of the lake. Figure (7-12) shows the variation of water levels upstream of the sea exit channel monitoring point.

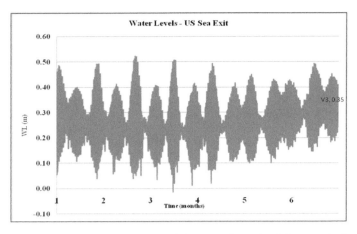

Figure (7-12): Water Levels calculated for the period of six months (Jan-July, 2006) at monitoring location US exit to sea

The velocity vectors show that the flow is affected mainly by the tidal forces, which is the main driving force in this case. Near the tidal boundary at the exit to the sea, the maximum flood and ebb tidal velocities are about 0.55 m/s as shown in Figure (7-13). Closer to the narrow link the flood and ebb velocities only reach 0.25 m/s. As we continue to move further away from the sea boundary into the mid region and northern and southern boundaries of the lake velocities tend to decrease from about 0.1 m/s to 0.05 m/s. For the monitoring location at the Edko drain inflow to the lake the maximum velocity reaches 0.45 m/sec.

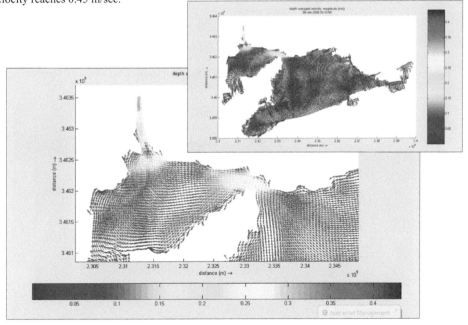

Figure (7-13): Depth averaged velocities simulation with flood tide and local drainage discharges in the Lake.

Case 2: Effect of Discharges, wind forces and Tidal Forces

In this simulation run the effect of the wind forces is specifically taken into account as one of the main physical parameters of the model. Hourly wind speed and direction values were used as physical inputs to the model with an average wind speed of 5 m/sec and a prevailing wind direction of NNW. The tide at the open boundary was defined by the same input data as in the first run of the model. In this simulation it is noticeable that the circulation differs strongly from the case with only the tide. The wind generates clockwise and anti-clockwise eddies inside the lake at different locations, as shown in Figures (7-14a, 7-14b), which are not dependent on the tide. The tide only affects a small area near the exit channel to the sea, and the circulation inside the lake is a result of the wind. The time varying NNW wind creates a strong southwest-east west set of currents.

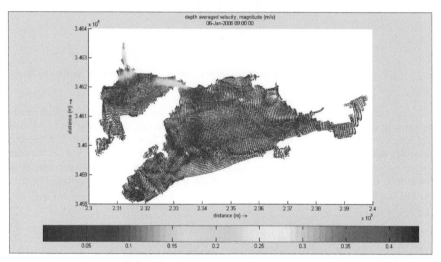

Figure (7-14 a): Depth averaged velocities simulation including tidal forces, wind forces and local drainage discharges in the Lake.

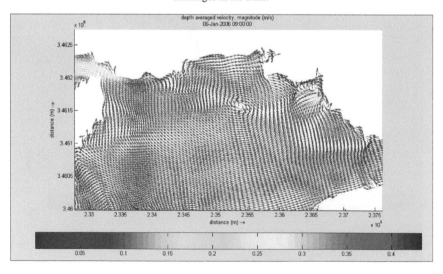

Figure (7-14b): Currents and Eddies formation after including effect of wind forces

The analysis of the velocity currents and wind speed shows a clear match between the directions of the current and the wind which indicates the main effect of the wind on the lake hydrodynamics. Figure (7-15) shows the analysis of the wind and velocity vectors.

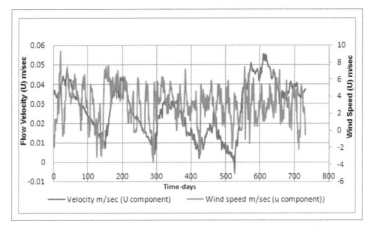

Figure (7-15): Comparison between wind and current velocities in the U direction

In this case the model outputs show that the drainage water inflows to the lake do not change the general circulation pattern inside the lake as generated by the wind: their influence is limited to the areas near their inlets.

Model Calibration
In order to calibrate the model, real data were used as boundary conditions and the results were compared with field measurements (that is, the water level at a fixed location upstream of the exit channel). The main calibration factor was the bed roughness coefficient. The hydrodynamic model was run for different roughnesses distributed over the grid. The roughness at the exit channel was increased from 0.05 to 0.09, and the main area of the lake different ranges of roughness coefficients were applied, namely 0.05-0.07-0.09. The roughness coefficients in the 2D model were determined to give a good representation of water levels measured at the exit channel to the sea, which has a value of 0.6 m. The roughness coefficient selected for the main water body was n=0.07m, while at the lake exit to the sea the roughness was increased to 0.09 to account for the exit bridge. Different roughness grids were tested and the target calibration parameter, namely the water level at the exit channel, was calculated with the selected roughness grid.

7.5. 2D WATER QUALITY MODEL OF SHALLOW LAKE SYSTEM

The water quality model for the shallow lake system has two main components. The first component is a 2D water quality model that consists of the main water quality parameters which are foreseen to be the main pollution indicators in the lake system. The second component is a eutrophication screening model. The development of these two models was determined by the complexity of the lake system and the interaction between the different lake components, namely the main water body, the vegetation and the fisheries. The aim of the 2D water quality model is to develop the temporal and spatial variations in the concentrations of main parameters in the lake. The eutrophication screening model is based on the modelling of CHL-a, which is considered a direct indicator of the eutrophication condition of the lake and is the main algae species existing in the lake water body. Figure (7-16) shows the water quality modelling and its components.

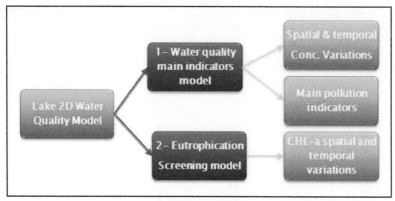

Figure (7-16): The 2D water quality modelling components of the shallow lake

The water quality and eutrophication models are coupled with the 2D hydrodynamic model developed for the water body. The Delft3D-WAQ module is used for the water quality modeling component coupled with the hydrodynamic module. Delft3D-WAQ is a 3-dimensional water quality modelling framework within the D3D modeling package. It solves the advection diffusion- reaction equation on a predefined computational grid for a wide range of model determinands. Delft3D-WAQ allows great flexibility in the substances to be modeled, as well as in the processes to be considered. It is important to note that Delft3D-WAQ is not a hydrodynamic model, so information on flow fields has to be provided by the hydrodynamic module.

A wide range of model substances is available in Delft3D-WAQ including:
• Conservative substances (salinity, CHL-aoride and up to five tracers)
• Decayable substances (up to five decayable tracers)
• Suspended sediment (up to three fractions)
• Nutrients (ammonia, nitrate, phosphate, silicate)
• Organic matter (subdivided in a carbon, nitrogen, phosphorus and silicon fraction)
• Dissolved oxygen
• BOD and COD (respectively Biological and Chemical Oxygen Demand)
• Algae
• Bacteria
• Heavy metals
• Organic micro-pollutants

Delft3D-WAQ allows the specification of an even wider range of physical, (bio)chemical and biological processes. These processes are stored in the so-called Process Library from which any subset of substances and processes can be selected. These processes include, for example: sedimentation and resuspension, reaeration of oxygen, Algae growth and mortality, mineralisation of organic substances, (De)nitrification, adsorption of heavy metals and Volatilisation of organic micro-pollutants.

The D3D-WAQ module is based on the mass conservation law. The model reproduces the mass balance of selected state variables. It does this for each computational cell. The mass transported by water flowing from one cell to the next serves as a negative term in the mass balance of the first computational cell and as a positive term of the second computational cell. By combining computational cells in one, two or three dimensions each water system can be represented. In particular the substances can be transported through the computational cells and hence through the water system. Delft3D-WAQ solves Eq. (7.4) for each computational cell and for each state variable for one step in time $(t+Dt)$,. Eq. (7.4) is a simplified representation of the advection-diffusion- reaction equation:

$$M_i^{t+\Delta t} = M_i^t + \Delta t \left[\frac{\Delta M}{\Delta t}\right]_{T_r} + \Delta t \left[\frac{\Delta M}{\Delta t}\right]_P + \Delta t \left[\frac{\Delta M}{\Delta T}\right]_S \qquad (7-4)$$

The mass balance has the following components:
- The mass at the beginning of a time step: $M_i^{t+\Delta t}$
- the mass at the end of a time step: M_i^t
- changes by transport: $\Delta t \left[\frac{\Delta M}{\Delta t}\right]_{T_r}$
- changes by physical, (bio)chemical or biological processes: $\Delta t \left[\frac{\Delta M}{\Delta t}\right]_P$
- changes by sources (e.g. waste loads, river discharges): $\Delta t \left[\frac{\Delta M}{\Delta T}\right]_S$

Changes by transport are a result of both advective and dispersive transport, that is, transport by flowing water and transport as a result of concentration differences respectively. The flow of water is usually derived from the Delft3D-FLOW hydrodynamic model (WL | Delft Hydraulics, 2005b). Dispersion in the vertical direction, which is important if the water column is stratified, is derived from Delft3D-FLOW as well. Dispersion in the horizontal direction is dependent on user input.

Changes by processes are due to physical processes such as reaeration and settling, (bio) chemical processes such as adsorption and denitrification, and biological processes such as primary production and predation on phytoplankton. Water quality processes convert one substance to another (such as the example of nitrification above).

Changes by sources are due to the addition of mass by waste loads and the extraction of mass by intakes. Mass entering over the model boundaries can also be considered as a source. The water flowing into or out of the modelled area over the model boundaries is derived from the Delft3D-FLOW hydrodynamic model.

Spatial Schematisation of the Water Quality Model
In order to model the transport of substances, the water system is divided in boxes as shown in Figure (7-17). The complete ensemble of all the boxes is called the 'grid' or 'schematisation'. In Delft3D-WAQ each box or grid cell is called a computational cell and is defined by its volume and its dimensions in one, two or three directions (Dx, Dy, Dz) depending on the nature of the schematisation (1D, 2D or 3D). Note that Dx, Dy and Dz do not have to be equal, so that the computational cell can have any rectangular shape. A computational cell can share surface areas with other computational cells, the atmosphere and the sediment or coastline.

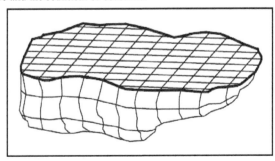

Figure (7-17): Division of a lake into small grids with a finite volume; a structured three dimensional grid

In Delft3D-WAQ each computational cell has a unique number ranging from 1 to N, where N is the total number of computational cells. Also, each surface area that is shared with another computational cell is identified by a unique number, ranging from 1 to Q, where Q is the total number of shared surface areas. Mass can be exchanged between the computational cells over these shared surface areas. Therefore, the shared surface areas are also referred to as *exchanges*. Delft3D-WAQ defines an exchange by the numbers of the two computational cells that share the surface area. For each cell, the volume, dimensions, surface area, and neighbouring computational cells (i.e. exchanges) are known. Thus, the water system is described by the individual computational cells, and through the exchanges it is known how the individual computational cells are interconnected. The basis for water quality modelling is the flow of water between the computational cells which is derived from the Delft3D-FLOW hydrodynamic model. Substances and water quality processes can be added to generate the water quality model.

7.5.1. Development of Main Water Quality Indicators Model

As explained in chapter 6, lake Edko, like most of the coastal lakes connected to drainage watersheds, is mainly used as a drainage sink to get rid of the continuous flow of drainage water coming from the catchment. This is in addition to other important uses of the lake, which include fishing inside the lake, preserving aquatic life, and sustaining fisheries around the lake area. For better understanding of the effect of different sources of pollution on the lake water quality and to assess the water quality, a water quality model is coupled with the developed hydrodynamic model to simulate the transport of different pollutants discharging to the water body of the lake through the two main drains.

The main pollutants reaching the lake are largely agricultural drainage water polluted with fertilizers and pesticides, untreated domestic wastewater, and waste loads from the fish ponds surrounding the water body and also from industrial wastes discharged to the upstream sub-catchments connected to the Edko main drain. The model aims at providing a better understanding of the fate of pollutants that enter the lake, and the spatial and temporal variations in the concentrations of these pollutants. The main objective of the model is to assess the seasonal variations of these pollutants in order to define the most critical indicators.

In discussing the practice of surface water resources engineering and water quality modeling, Rafailidis (1994) reports that the following determinants are of importance: Carbonaceous Biochemical Oxygen Demand (BOD), Dissolved Oxygen (DO), Ammoniac compounds and coliform Bacteria. BOD indicates the overall organic pollution of the water, and (DO) shows whether the aquatic life may be sustained there. The nutrient concentration (indicated by forms of nitrogen) gives the potential for eutrophication. Coliform counts indicate the danger of disease for humans using the water. The temperature, salinity and total suspended matter are also considered of importance in the modelling.

The parameters selected for the water quality model were based on the main water use of the lake, namely, the aquatic life and the fisheries. Also the existing pollution problems in the lake area indicated that the main problem there is eutrophication in addition to organic pollution from the untreated human waste and fisheries waste dumped into the lake through the two main drains. Therefore, the selected parameters are divided into three groups for the study of these problems. The first group consists of general variables including: Temperature, Salinity, dissolved oxygen and total suspended matter. The second group includes the nutrient variables: nitrogen compounds (ammonia $NH4$ and nitrates $NO3$) and phosphorous compounds (orthophosphate $PO4$). The third group is formed from the organic compounds variables and it includes: the biological oxygen demand BOD and the chemical oxygen demand COD.

The Processes Involved in the Water Quality Model

In Delft3D-WAQ module the constituents of a water system are divided into functional groups. Figure (7-18) shows these groups and the links between them. A functional group includes one or more substances that display similar physical and/or (bio)chemical behavior in a water system. For example, the nutrients: nitrate, ammonium, phosphate and silicon, are a functional group as they are required for primary production. Functional groups can interact with each other directly or indirectly. The highlighted substance groups in Figure (7-18) indicate the groups including the selected variables for the developed water quality model. It is important to mention that the substances and water quality processes to be modeled are created or selected with the PLCT (Processes Library Configuration Tool). The PLCT is a component of the Delft3D-WAQ module. This library allows us to select the following inputs to the model: State variables (called substances in Delft3D-WAQ), water quality processes, editable process parameters and output variables.

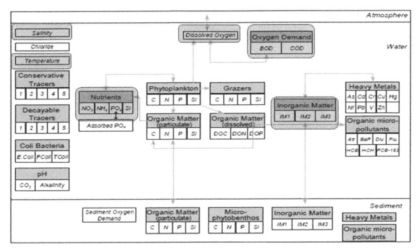

Figure (7-18): General overview of substances included in Delft3D-WAQ.
(After Delft3D-WAQ manual, 2005)

The basic steps in water quality modelling applying the Delft3D –WAQ module could be sumarised in the following:

1. Selection of the result from the hydrodynamic simulation and make it suitable for application in the water quality simulation (coupling Process).
2. Definition of the substances and water quality processes to be included in the model.
3. Definition the water quality simulation properties using the outcome of the first and second step.
4. Definition of initial conditions, boundary conditions, waste loads, simulation time, output variables and identification of monitoring points
5. Run the simulation and check the output.
6. Calibrate and verify the model.

Figure (7-19) shows the data flow diagram for the Delft3D-WAQ module

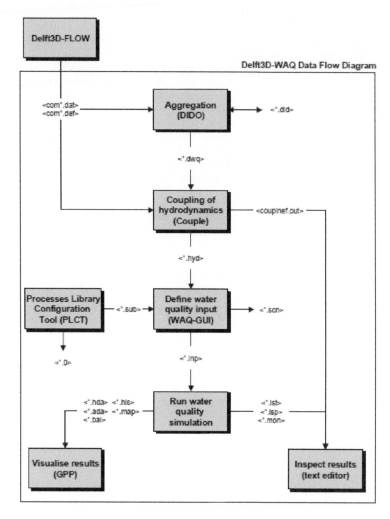

Figure (7-19): Overview of the modules and data flow diagram in Delft3D-WAQ
(After Delft3D-WAQ manual, 2005)

Model Inputs Data Groups

The water quality main model is developed after coupling it with the 2D hydrodynamic model of the lake. *The coupling* is the first step in building the water quality model. The coupling process allows Delft3D-WAQ module to make use of the hydrodynamic conditions and parameters (velocities, water elevations, density, salinity, vertical eddy viscosity and vertical eddy diffusivity) calculated in the Delft3D-FLOW module.

The development of the substances list or the list of parameters to be modeled is done through using the PLCT library as mentioned above. The same list includes all the processes associated with each parameter. The selected substances groups and parameters are described in Table (7-1).

Table (7-1): Delft3D-WAQ selected model substances groups, parameters and associated processes

Substance Group	Selected Model parameters	Associated Processes
General	Continuity, water temperature, Salinity	Temperature and heat exchange
Oxygen-BOD	BOD, COD, DO	Mineralisation BOD and COD, Sedimentation COD, Sedimentation COD, reaereation of Oxygen.
Suspended Matter	Inorganic matter IM1 (TSM)	Sedimentation IM1, Resuspension 1st inorganic matter
Eutrophication	Ammonium (NH4), Nitrate (NH3), Ortho-Phosphate (PO4)	Nitrification of ammonium, Denitrification of nitrates

The initial conditions for the model were selected to be zero concentrations for all modeled parameters except the continuity parameter which checks the mass balance of the model. It was set to 1 g/m³, so that the model can run to predict the pollutants concentration in the lake at different selected monitoring points based on the boundary conditions concentration of the inflow from the two main drains. The model simulation period was selected as the same period for the hydrodynamic modeling, namely, from the beginning of January 2006 till the end of July 2006. *The time step* selected for the water quality model was set to 1 hour. *The process parameters* for some selected modelled substances are set to their default values and they are by default constant in time and space. However, process parameters can vary in time and/or space. Initially process parameters will have the default value that is taken from the PLCT. These values may be changed in the *Process parameters* data group.

Model Boundary Conditions and Observation Locations

The boundary sections for the water quality model are selected to be the same boundary sections for the hydrodynamic model at the locations of the main drains outlets to the lake, where all discharges enters the lake. Concentrations for different modeled parameters are defined as time varying boundary conditions at the two sections for the Edko Drain outlet and Barseek drain outlet. The concentrations used at the boundaries are time series average monthly concentrations for the modeling period. This time series is developed by linear interpolation between different time breakpoints. The observation history points for monitoring model outputs are defined to be the same locations for the field water quality sampling locations for the evaluation and calibration processes. Figure (7-20) shows the schematization of the open boundaries and observation points locations.

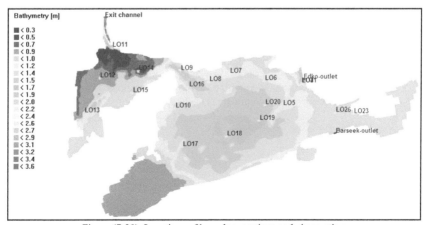

Figure (7-20): Locations of boundary sections and observation points for water quality model of lake Edko.

Mass Balance Verification of the Water Quality Model

A special type of conservative tracer in Delft3d –WAQ is a parameter called 'Continuity'. It has no physical or chemical meaning. Instead it is used to establish the numerical correctness and stability of the simulation. By assigning a concentration of 1 g/m3 to all water sources (: initial condition, boundary condition and discharges) the Continuity concentration should remain 1 g/m3 during the whole simulation, as there are no processes that dilute or concentrate it and all water has a concentration of 1 g/m3.To verify the mass balance of the developed water quality model and the simulation stability the continuity parameter was simulated for the whole modelling period. The variation of the results at selected locations in lake shows acceptable range between (0.999 and 1.005). Figure (7-21) shows the results of mass balance simulation by the model at different monitoring locations.

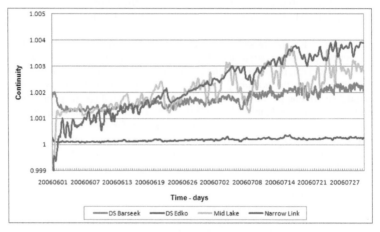

Figure (7-21) shows the results of mass balance simulation by the model at different monitoring locations within the lake.

Model Results

The main water quality model includes the first three substances groups mentioned above (general group including temperature and salinity), (oxygen group) and (suspended matter group). The following section shows the results of the model during field measurements (June27[th]-29th, 2006).

The predicted values of temperature are in the range of (25-28 °C) and this is the same range of measured temperatures at the same locations during field measurements. The results of salinity predictions showed a close matching with the field measurements, where the ranges of salinity are between (1.4-1.37 g/kg). Figure (7-22) shows the water Temperature levels at different locations in the lake. Figure (7-23) shows the salinity variations at the same locations. The results from modelling of the oxygen group parameters are shown in figures (7-24 a,b,c).

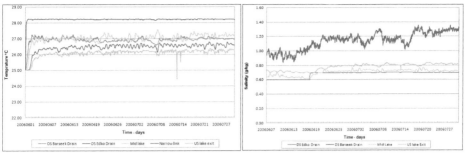

Figure (7-22): Modelled water temperature at different measuring locations

Figure (7-23): Modelled water temperature at different measuring locations

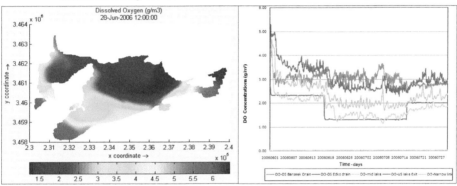

Figure (7-24-a): Model results of Dissolved Oxygen concentrations at different measuring locations

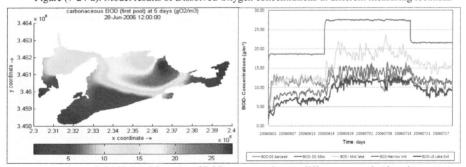

Figure (7-24-b): Model results of BOD concentrations at different measuring locations

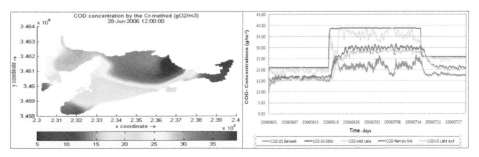

Figure (7-24-c): Model results of COD concentrations at different measuring locations

The TSM results show a spatial and temporal variation. In general there is an adequate agreement between simulations and observations, although at some locations the field observations showed high concentrations than the model results. The concentration varies at different locations as shown in shown in figures (7-25 a,b). At the drain exit into the lake TSM has an average of 38 mg/l while at the middle part of the lake average TSM is 34 mg/l. It is noticed that at the Exit of Barseek drain the TSM have high concentration with an average of 46 mg/l.

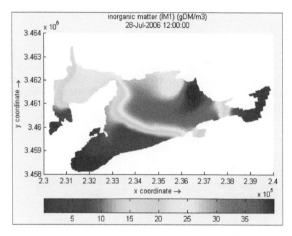

Figure (7-25-a): Modelled TSM map (June 28th, 2006)

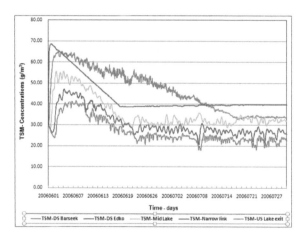

Figure (7-25-b): Modelled TSM at mid of lake (June-July, 2006)

Model Calibration

A successful model application requires a model calibration that compares simulated results with measured lake conditions. This section discusses the model calibration procedures and details. In general the model results are tested or calibrated against field measurements. This field data must be different from the data set used during the construction of the original model (Thomann, 1982; Thomann and Mueller 1987). The objective of model calibration is to adjust the input parameters so that there will be closer agreement between the simulated values and observed data (Ambrose, 1992; Bierman, 1986). There are several methodologies and techniques applied for water quality models calibration. Dilk's *et al.,*(1990) have calibrated a DO model by modifying the nitrogen rate, COD and BOD deoxygenation rate, and reaereation rate. Ambrose (1992) has calibrated an estuary water quality model by adjusting dispersion values in transport processes and reaction rate coefficients in water quality interactions. Additionally lung and Larson (1995) have used low flow conditions to calibrate a water quality model. Another different methodology was applied by (Masato *et al.,* 2002), where the water quality model calibration was regarded in their study as an optimisation problem to minimize the discrepancy between the observed and calculated results; global optimisation was used.

In this study the water quality model calibration is done on different levels and by applying different techniques. The first level is presented in this chapter. Here the conventional water quality parameters or oxygen group (DO, COD, BOD and Nutrients group NH4 and NO3) and their associated model process parameters are selected for calibration. In particular, the process parameters are adjusted for this level of calibration. In the next chapter, further calibration procedures based on the application of remote sensing techniques are used for enhancing the model performance for specific parameters such as TSM and CHL-a. The first level calibration was based on a comparison between simulations and measurements and the calculation of the Mean Relative Error (MRE) and the Root Mean Square Error (RMSE) to examine the performance of the model. Due to the scarcity in data inside the lake the comparison and calibration was done by comparing concentrations on a spatial basis between predicted and measured values. The following section presents the different calibrated parameters and the calibration results.

Calibration of the Oxygen Group Parameters

The first level of model calibration was carried out by visual comparison of simulations and measurements in graphs together with the calculation of the statistical error values such as Mean Relative Error (MRE); the overall performance of the model was examined as well. The output variables of the model such as DO, COD, BOD with respect to the observations in Lake Edko during field survey were plotted in graphs to make comparisons, which were used to check how the simulations fit the observations. Besides, MRE was used to quantify the agreement of the model, by dividing the residuals by the observed values. In this study the calculation of RE and MRE was based on Equations (6-5) and (6-6):

$$RE = \frac{Csim - Cobs}{Cobs} \times 100 \qquad (7\text{-}5)$$

$$MRE = \text{Sum} \left| RE \right| \div n \qquad (7\text{-}6)$$

Here Csim and Cobs are the simulated and observed values respectively, and n is the number of cases. The MRE denotes the mean relative difference between simulations and observations. Other statistical methods for error calculations were used such as Mean Error (ME), Mean Absolute Error (MAE) and the Root Mean Square Error (RMSE). The oxygen group calibration parameters are shown in the following Table (7-2):

Table (7-2): Calibration process parameters for water quality model

Calibration parameters	run1 (default)	run2	run3
Reareation transfer coeff (m.d^{-1})	1	2	3
Decay rate BOD (d^{-1})	0.3	0.15	0.08
Decay rate COD (d^{-1})	0.05	0.025	0.01
Temp coeff. decay rate BOD	1.04	1.04	1.04
Temp coeff. decay rate COD	1.02	1.02	1.02

Calibration was done for two sets of model parameters values. It is important to note that the temperature is an important water quality parameter that affects almost every water quality process. Therefore the temperature coefficients for the decay rates of oxygen group parameters are also used in the calibration process. Results of calibration of the oxygen group are shown in the following figures (7-26) to (7-28):

1. Dissolved Oxygen

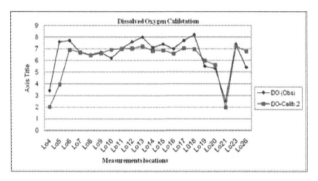

Figure (7-26): The calibrated result of dissolved Oxygen (DO).

It is noted that at the entrance of the main drain to the lake (LO4) the DO has the lowest values; in general the DO measurements are close to the simulated results with an RME value of 11.78%.

2. BOD $_5$

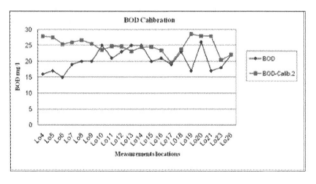

Figure (7-27): The calibrated result of BOD5

The BOD values show a difference between simulated and observed values at locations near the drain inlets to the lake, where the model over estimated the values of BOD. Although the trends are quite similar, the values are simulated with a relative error of 48%. This could be due to the low velocity distributions at these locations around the lake edges; but the overall MRE for all measurement locations is within an acceptable range of 18.29%.

3. COD

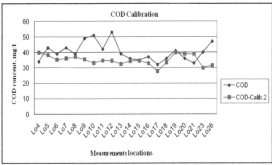

Figure (7-28): The calibrated result of COD

The simulated COD results are very close to the measured values at most locations within the lake, but with noticable different values and trends at locations LO9, LO10 and LO13, which could be due to the inaccuracy of sampling analysis. This is because the trend of the model is in line with the trend of measurements in general, and the MRE values is around 15.72 %.

Analysis of Results of the Calibrated Model

In general, we can conclude that the calibration of the oxygen group parameters shows that there is close agreement between the simulated values and observed data, with an acceptable range of error since the model performance is very much dependant on the sufficiency and accuracy of the measured data. But since the modelling of water quality parameters is a complicated process that involves a variety of uncertainties we can accept the relative errors at the investigated locations. Table (7-3) shows the different values of RE and MRE for the modelled parameters at this level. The overall calculated statistical parameters for the calibrated model are given in table (7-4).

Table (7-3): Relative error for the calibrated model parameters

Location / Parameter	DO (RE %)	COD (RE %)	BOD (RE %)
Lo4	17.32	41.18	42.69
Lo5	10.77	48.29	38.27
Lo6	9.36	10.13	40.81
Lo7	15.77	0.15	26.84
Lo8	4.74	0.92	24.90
Lo9	27.35	1.04	21.38
Lo10	35.00	11.77	6.02
Lo11	17.33	0.00	15.25
Lo12	34.72	7.50	6.84
Lo13	16.85	9.88	8.08
Lo14	4.61	3.80	2.08
Lo15	1.54	7.16	18.60
Lo16	11.22	5.86	10.22
Lo17	13.66	8.44	2.91
LO18	7.31	15.00	2.83
Lo19	2.83	8.91	40.50
Lo20	8.14	6.04	6.88
Lo21	17.64	22.00	39.00
Lo23	25.40	2.16	11.59
Lo26	32.94	25.56	0.14
MRE	15.72%	11.79%	18.29%

Table (7-4): The overall calculated statistical parameters for the calibrated model

Statistical Parameter	DO	COD	BOD
MRE	15.72%	11.79%	18.29%
ME	0.407	5.296	-4.423
MAE	0.709	6.760	4.803
RMSE	1.062	8.616	6.300

Discussion on Calibration Results

Due to the lack of temporal data sets at the measuring locations in the lake the calibration was limited to comparing the model results with the observed data at the different locations. The calibration of the main oxygen parameters could be considered to have acceptable values according to the corresponding statistical parameters values. The model results and calculations are in reasonable agreement with the measured concentrations. The differences for these parameters could be attributed to the uncertainty in the model input data due to errors in monthly measurements; also the time frequency of measurements with a month time difference is considered a very big time gap for water quality monitoring. This is in addition to the expected uncertainty with the samples analysis. Therefore to improve the model performance we need to have more extensive monitoring and investigation of input data at the boundaries.

Sensitivity Analysis

During the calibration of the model the different process parameters were within their theoretical ranges. This observation was used in a sensitivity analysis to change these values and check the variations of specific parameters in the water quality calculations. The sensitivity analysis was done by varying the following parameters; decay rate of BOD (RcBOD), nitrification rate (RcNit20), then the group of temperature coefficient of BOD decay (TcBOD) , Nitrification (TCNit) and sediment oxygen demand factor (fSOD); a third group includes the sedimentation velocity of BOD. The following section shows the results of this analysis. The analysis is divided into three categories; high effect parameters, less effect parameters and low effect parameters. A range between the maximum and minimum values of parameters is used based on values published in the literature. Three locations at different zones of the lake are selected for the sensitivity analysis of model parameters on the DO concentration.

Sensitive Parameters (high effect)

These are the first group of model parameters including the decay rates of BOD (RcBOD), the temperature coefficient for BOD decay (TcBOD), and the natural temperature of the water. Figure (7-29) shows that the decay rate of BOD is a sensitive parameter that affects the predicted concentration of both DO and BOD; the same is true for the decay temperature coefficient, where the analysis shows a considerable variation in the concentration of DO when changing the parameter within the range of 1.03 to 1.05. Figure (7-30) shows the variations within DO concentrations with different values of (TcBOD). In Figure (7-31) the effect of TcNit is shown on the concentration of DO. Table (7-5) shows the sensitivity analysis of these parameters. It is concluded from the sensitivity analysis of the first group of parameters that the main sensitive parameters are oxygen groupparameters; the decay rates of BOD (RcBOD) and the temperature coefficient for BOD decay (TcBOD), and nutrients group parameters; nitrification decay rate (Rc-Nit) and nitrification temperature coefficient (Tc-Nit).

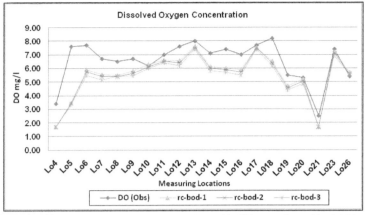

Figure (7-29): variations within DO concentrations with different values of (rc-BOD)

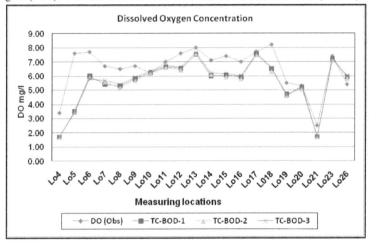

Figure (7-30): variations within DO concentrations with different values of (TcBOD)

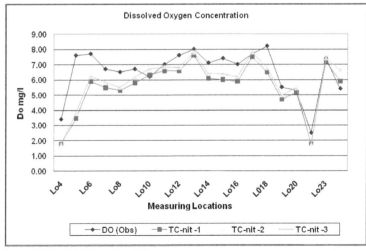

Figure (7-31): variations within DO concentrations with different values of (Tc-Nit)

Table (7-5): Statistical evaluation of sensitivity analysis

Parameter Values	Statistical Methods			
	ME	MAE	RMSE	R^2
RcBOD;(1/d): 0.15	0.407	0.709	1.062	0.63
RcBOD;(1/d): 0.17	1.086	1.104	1.426	0.56
RcBOD;(1/d): 0.2	1.224	1.239	1.541	0.65
RcNit20;gN/m3/d):0.1	0.407	0.709	1.062	0.63
RcNit20;gN/m3/d):0.5	0.407	0.709	1.062	0.63
RcNit20;gN/m3/d):1.0	0.407	0.709	1.062	0.63
TcBOD: 1.03	1.042	1.082	1.401	0.66
TcBOD: 1.04	0.982	1.037	1.357	0.660
TcBOD: 1.05	0.909	0.984	1.338	0.64
TCNit20;(1/d): 1.05	0.969	1.029	1.357	0.647
TCNit20;(1/d): 1.07	0.684	0.880	1.198	0.648
TCNit20;(1/d): 1.1	0.684	0.880	1.198	0.648

The overall statistical analysis indicates that the decay rate of BOD is the most sensitive parameter to the modelled DO concentration. The relation is an inverse relation, i.e. increasing the decay rate parameter decreases the DO concentration. The temperature coefficient is also of significantly sensitive in the model. It should be highlighted here that the calibration of the TSM (and CHL-a) was done in this research study based on the application of remote sensing techniques and methodologies as presented in chapter8. An important objective in this research was to test and prove the capabilities of remote sensing in filling the gaps in data scarce environments, and overcoming the complexity of the calibration of both modelled parameters when there are practical limitations of data availability. Chapter 8 will involve the detailed calibration procedures of TSM as one of the parameters of both the basic model and the eutrophication screening model.

7.5.2. Eutrophication Screening Model of the Shallow Lake System

A good and reliable understanding of the lake ecosystem is essential for economical and ecologically feasible and sustainable lake management. Most managers are very interested in models which can be powerful tools to simulate the changed water quality of lakes due to eutrophication and to test various management options to restore the lake conditions. Ever since it emerged in the early 1970s, eutrophication modeling has been considered a step forward from the hydrodynamic models because it incorporates chemical and biological processes as well as environmental management aspects into the transport processes. Since then numerous eutrophication models have been developed to simulate different ecological and hydrodynamic processes with a wide range of different complexity in both processes. The eutrophication model is considered a complex model with respect to its structure and its data requirements, as it includes both chemical and biological sub-processes which interact in several meta processes and these processes are linked directly with the hydrodynamic results.

For the case of Lake Edko as an example of a shallow lake system under eutrophic stressing conditions, a simple eutrophication screening model is developed as an initial step for managing the lake system. The detailed processes of eutrophication were not all included because the scope of this study is not the ecological modeling of lakes but to make use of the hydrodynamic model and couple it with the eutrophication model for a better understanding of the conditions of the lake.

The eutrophication model developed for the Lake includes the following parameters of interest: total suspended matter (TSM), nutrients group including ammonia (NH4) and Nitrates (NO3) and CHL-a. These parameters were selected as the main indicators for the eutrophication condition of the lake.

Model development and assumptions

The model is developed using the phytoplankton module D3D-ECO:BLOOM module which is a component under the Ecological module of D3D. In respect of the level of detail of the ecological modules, the phytoplankton module (BLOOM) is the most extensive as it includes several functional groups and types. BLOOM is a multi-species algae model, based on an optimisation technique that distributes the available resources in terms of nutrients and light among the algae species (WL, 1991 and 1992; Los and Brinkman, 1988). BLOOM optimises the species composition to obtain the overall maximum growth rate under the given conditions. A large number of groups and/or species of algae and even different phenotypes within one species can be considered. BLOOM distinguishes between three phenotypes: under nitrogen limiting conditions, under phosphorus limiting conditions and under light limiting conditions. In general, the availability and accuracy of the data needed to determine the values of the model coefficients as well as the data for model validation limit the reliability of modeling results. Naturally, this is true for algae modelling as well.

The main objective of the model is to have a better understanding of the nutrients condition in the lake which is reflected by the CHL-a prediction results, The CHL-a is a main output from the model that is used as an indicator of eutrophication. Modeling of CHL-a implies the modeling of the main nutrients group (NH4, NO3, PO4) Total suspended matter (TSM) and the existing algae species in the lake.

The general mass balance for phytoplankton (in the water column) is given in the following equation for BLOOM module:

$$\left(\frac{\Delta \text{Phytoplankton}}{\Delta t}\right) = \text{Loads} + \text{Transport} + \text{Settling} + \text{Gross primary production}$$
$$- \text{Respiration} - \text{Mortality} - \text{Grazing}$$

(6-7)

Model Input Parameters

An important parameter to set up the BLOOM module is the definition of the dominant algal species in the water body. The modeled algae species in the case of Lake Edko are diatoms and green algae as these are the dominant species in the lake water according to (Fathi *et al*, 2001). The model was developed based on the work done by LOS et al, 2007. The following assumptions were used in developing the BLOOM model for Lake Edko. It is assumed that 1 g CHL-aorophyll-a corresponds to 7.5 g N and 0.75 g P in phytoplankton. This corresponds to a g C/CHL-a ratio of 50 and N/C and P/C ratios of 0.15 and 0.015, respectively. According to Los et al, 2007, since the primary production is strongly influenced by light availability, and can even become limited if there is too little light, the calculation of light conditions in water is an important process in the model. The light limitation function can be based on daily average and depth average conditions. This function is associated with the critical ambient extinction coefficient which is species specific. The coefficient is derived from imposed tables that relate production efficiency to ambient light intensity (irradiation). Growth inhibition may be included in these tables if the radiation is larger than the optimal radiation. These tables are part of the BLOOM data base. The main stoichiometric coefficients values used in the model are shown in Table (7-6). The model inputs are divided into the following sets:
- Substances file including all modeled, processes and results parameters
- Boundary conditions
- Initial conditions
- Processes parameters
- Stoichiometric coefficients

Table (7-6): Stoichiometric coefficients for Eutrophication model (*All ratios are in(mg/mg)*)

	Dry matter	Nitrogen/Carbon	Phosphorus/Carbo	Silicon/Carbon	Chlfa/Carbon
Fresh DIATOMS energy type	3	0.21	0.018	0.66	0.04
Fresh DIATOMS P/Si type	2.5	0.188	0.0113	0.55	0.025
GREENS energy type	2.5	0.275	0.0238	0.0018	0.033
GREENS nitrogen type	2.5	0.175	0.015	0.0018	0.025
GREENS phosphorus type	2.5	0.2	0.0125	0.0018	0.025

All model inputs were based on the database collected for water quality nutrients parameters NH4, NO3 and PO4, in addition to Si and TSM concentrations. The model was run for the same period of the water quality model (six months) with a time step of 6 hours. The modeling results are shown in the following section.

Model Results and Discussion

The first runs of the model showed in general high concentrations of CHL-a that range from 20 mg/m^3 to 100 mg/m^3. These concentration ranges were considered to be initially accepted, referring to the field measurements taken during the work survey in 2005 when the average concentration was 60 mg/m^3. Also, the model results are comparable to those in the literature, which shows that the average values of CHL-a are the order of 60 mg/m^3 (Siam et al, 2000). The detailed analysis of the model results indicates thaat the concentration ranges between 78-88 mg/m^3 at the exit of Edko drain into the lake and between 50-80 mg/m^3 in the middle area of the lake, as shown in Figures (7-32) and (7-33). In the western area of the lake the concentrations are in range between 25-100 mg/m^3 as shown in Figure (7-34).

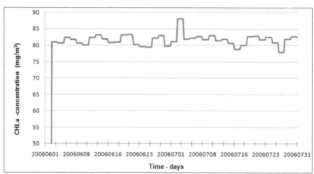

Figure (7-32): Concentrations of CHL-a at Edko drain exit to the lake

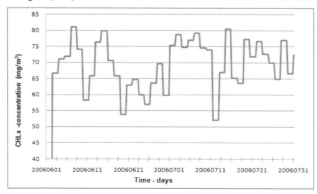

Figure (7-33): Concentrations of CHL-a at mid zone of Lake Edko

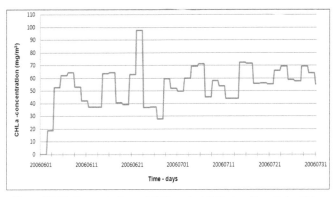

Figure (7-34): Concentrations of CHL-a at mid-Western Zone Lake Edko

The concentrations of ammonia NH4, nitrates NO3 and Phosphorous PO4 are also shown in the Figure (7-35).

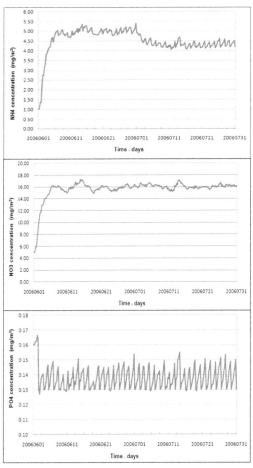

Figure (7-35): concentrations of NH4, NO3 and PO4 mid lake

161

The spatial distribution of the CHL-a concentration varies with time according to the variation of the hydrodynamics effects due to the wind. The CHL-a map for 28th of June is shown in Figure (7-36). The figure illustrates high concentrations in the middle area of the lake. This date was selected for the calibration procedures using remote sensing data.

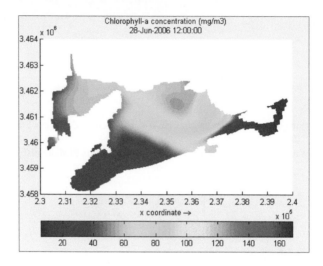

Figure (7-36): CHL-a map on 28th of June 2006

The calibration of eutrophication models is considered a complicated process due to the involvement of several parameters and processes. In the eutrophication screening model, CHL-a is considered the main indicator parameter for eutrophication. Due to the lack of CHL-a monitoring data and the ecological model parameters, traditional calibration procedures are not applicable in this case; therefore calibrating the CHL-a model will be mainly based on remote sensing data as highlightes earleir. The calibration of the model is described in chapter 8. The developed BLOOM phytoplankton module was applied to establish the relations between phytoplankton biomass (e.g. CHL-a) and physico- chemical quality elements and pressures (e.g. nutrients and light conditions).

8. APPLICATION OF REMOTE SENSING FOR ADJUSTMENT AND CALIBRATION OF WATER QUALITY MODEL

Conventional measurements of water quality parameters, especially those related to eutrophication, namely TSM and CHL-a, require in situ sampling and expensive and time-consuming laboratory work. Due to these limitations, the sampling size often cannot be large enough to cover the entire water body. Therefore the difficulty of synoptic and successive water quality sampling becomes a barrier to water quality monitoring and forecasting (Shafique, 2001). The problem of sampling has its own implications for any further application to water quality assessment such as with mathematical modeling and the calibration of water quality models. Therefore remote sensing is introduced as a tool to overcome and complement the lack of field data that is needed in the calibration process. This chapter shows different methodologies to calibrate the models using different types of remote sensing data and different levels of calibration. It also highlights the advantages and disadvantages of these methods.

8.1. INTRODUCTION

An important reason for integrating remote sensing into the water quality management information system is the scarcity of field monitoring measurements due to the absence of water monitoring stations in the Edko lake area. To overcome the lack of calibration data, remote sensing provides an alternative means of water quality monitoring for a range of temporal and spatial scales (Yang et al 2000).Therefore this chapter focuses on the development of different procedures based on remote sensing techniques for the validation of the hydrodynamic model and the calibration and enhancement of the water quality mathematical model developed for Lake Edko. The use of remote sensing data driven from both satellite images and spectral in-situ measurements can assist in developing spatial and temporal data sets that can be directly used for model calibration and enhancement. In this research, the complexity of the physical and ecological properties of the lake system was sufficient reason to explore different methodologies and procedures for model calibration using remote sensing. Here, the role of remote sensing is considered of great importance in order to compensate for missing in situ measurements.

The calibration methodologies used are categorized as qualitative and quantitative. The different techniques of remote sensing for estimating water quality parameters have their own limiting factors. In this study calibration was based on two parameters: total suspended sediments and CHL-aorophyll-a, Many satellite sensors are potentially suitable for estimating the concentration of these parameters. The basis for sensor comparison and selection is the spectral, spatial and temporal resolution, in addition to factors related to the quality of selected images and cloud coverage, which may affect the selection of images. As mentioned in Chapter 5, two types of sensor data are used for the calibration process: the Moderate-resolution Imaging Spectroradiometer (MODIS) and the SPOT-5, characteristics of these data sets are given in Annex (A1).

8.2. DESCRIPTION OF THE APPLIED REMOTE SENSING METHODS

Comprehensive research on the quantification of water quality parameters reflectance spectra began in the early 1970's. Remote sensing of freshwaters can basically be done through two different approaches or through a combination of the two: The empirical or (statistical approach) and the analytical or bio-optical modeling approach (Gordon and Morel, 1983).

8.2.1. Empirical (Statistical) Approach

The empirical approach is based on the calculation of a statistical relation between the water constituent concentrations and the radiance measured by the sensor. Normally in the empirical approach, remote sensing data is related by regression analysis to *the lake in-situ* measurements of water quality parameters. This approach needs extensive field work and logistics since samples have to be collected from the water body simultaneously or near simultaneously with the overpass of the sensor, which in practice is very difficult to achieve. A review of the literature on empirical algorithms for estimating water quality parameters shows a vast variety of algorithms proposed. They start from a simple linear regression between reflectance and water constituent concentrations to non-linear multiple regressions between a combination of band ratio(s) and the concentrations. The advantage of using the empirical approach is that the algorithms are straightforward and easy to use. The disadvantages are that false results may occur while using this method, because a causal relationship does not necessarily exist between the parameters studied (Hogenboom and Dekker, 1999).

8.2.2. Analytical Approach (Bio-optical Modelling)

In the analytical approach, an in-water optical model is developed to interpret the remote sensing data and retrieve the water quality concentrations. These models use the inherent optical properties or concentrations of different optically active substances as input, and give estimates of the radiance reflectance as the output. The models are then inverted to determine the concentrations of the optically active water quality parameters from a spectrum of radiance reflectance (Jupp, Kirk et al. 1994; Keller, Keller et al. 1998; Pierson 1998). Several models for coastal and inland waters were investigated by Gordon et al. (1975). They are similar to a solution of the radiative transfer equation in which volume reflectance is expressed as a function of absorption and backscattering coefficients of the water constituents. Bio-optical models have been proven to be suitable for interpreting remote sensing measurements of lakes if the water quality parameters are related carefully to the inherent optical properties of water (IOP's).

One of the main models developed is that of Dekker (1993), which was applied to Dutch lakes datsets. The main water quality parameters used in this model were: the colored dissolved organic matter (CDOM), CHL-aorophyll (CHL-A) and total suspended matter (TSM). The model equations are as follows:

$$R(0-) = f \frac{b_b}{a + b_b} \tag{8-1}$$

$$b_b = b_w \times B_w + b_{tsm}^* \times B_{tsm} \times TSM + b_{chl}^* \times B_{chl} \times CHL \tag{8-2}$$

$$a = a_w + a_{tsm}^* \times a_{cdom}^* \times CDOM + a_{chl}^* \times CHL \tag{8-3}$$

where:

R(0-) is the subsurface irradiance reflectance;
f is a proportionality factor related to the illumination condition and viewing geometry;
b_b is the total backscattering coefficient;
b^*_{tsm} and b^*_{CHL-a} are the specific scattering coefficients of TSM and CHL-A respectively;
B_w, B_{tsm}, B_{CHL-a} are the probabilities that light will scatter back to the sensor from a given water constituent;
a is the total absorption coefficient ;
a^*_{tsm}, a^*_{cdom}, a^*_{CHL-a} are specific absorption coefficients of TSM, CDOM and CHL-A respectively;
a_w , b_w are absorption and scattering coefficients of pure water;
TSM, CDOM and *CHL-A* are concentrations of water constituents: TSM, CDOM and CHL-A respectively.

The parameters in the above equation are the specific absorption and scattering coefficients of the water constituents, commonly known as Specific Inherent Optical Properties (SIOPs). Essentially, these parameters are the absorption and backscattering per unit of water constituents, and they change from one type of water body to another. The main advantage of this approach is that once the optical properties of studied water bodies are identified, the model can be applied to any remote sensing image irrespective of the time of its acquisition. A clear disadvantage is that the model uses various input parameters, which are often not available (that is, they are difficult to measure in the field or from field samples) or are very much affected by the bottom reflectance and submerged vegetation. The accuracy of the model is based very much on the sensitivity to these different parameters.

In this study there were not enough measurements in the lake to define the SIOP's of the water body due to practical limitations. However, both methodologies were tested and applied in this study with certain limitations to develop the concentrations of TSM and CHL-a for Lake Edko. The aim was to use them in the calibration procedures of the water quality and eutrophication models.

8.3. LINKING REMOTE SENSING WITH WATER QUALITY AND EUTROPHICATION MODELS: CALIBRATION PROCEDURES

Water quality modelling is associated with a recognised level of uncertainty, which arises from different sources including the measurements, model inputs, model parameters and the extensive complicated interactive processes that involve different water quality parameters. Therefore these uncertainties are also expected to strongly influence the calibration process of the water quality models. The problem becomes more complicated when the model being calibrated is a *data scarce model,* that is, when the data sets required for model calibration are insufficient for the spatial and temporal requirements needed to perform a standard model calibration, as shown in Chapter 7. In the present chapter, remote sensing is introduced as a tool for calibrating water quality and eutrophication models for the parameters TSM and CHL-a. Different levels of model calibration could be done using different levels of water quality data sets generated from remote sensing data. Remote sensing can be seen as a complementary tool in the water quality modelling process for similar situations to Lake Edko, as this study proves, where the lack of in situ measurements is a common feature due to economic and technical constraints.

In the following sections, a detailed explanation of the advantages of using remote sensing data for calibration is highlighted, and the different types of used satellite images and remotely sensed data are presented. The different methodologies for calibrating the water quality and eutrophication models are explained and the advantages and limitations of each used methodology are discussed.

8.3.1. Advantages of Calibration Using Remote Sensing Data

Most water quality models have a complex structure and include a number of processes such that the simulation results are most of the time linked with high uncertainty. However, most of the model parameters that greatly affect the results should be assigned in advance by referring to typical values found in the literature. This is because field data work usually misses these modeling parameters. Consequently, the error caused by the uncertainty in the coefficients greatly decreases the model's reliability (Canale and Seo, 1996).

Therefore, developing a suitable calibration procedure for these model parameters is important before the model can be implemented for a real aquatic system. Model calibration is an essential process to test and tune a model by comparing the simulated results and field data. The simplest method is the parameter tuning and trial-and-error, which was introduced in Chapter 7 due to lack of temporal field measurements. By minimizing the difference between simulated results and field data, the modeller can set up a water quality model with one set of rational parameters for a specific water body. However, this guesswork is time-consuming and relies considerably on the user's experience. The reasons why statistical methods are not commonly used in model calibration are the complexity of the models and the limitation of a large field data set (Henderson-Sellers and Davies, 1991). Moreover, field data are not always enough in both spatially and temporally. Where there are not enough data sets for calibration the statistical measures of goodness of fit will not be possibly applicable, therefore remote sensing technique could provide missing spatial and temporal data sets, which therefore could be used for calibration.

Early studies that took on board the use of remote sensing data in integration with mathematical models, showed promising results. Several research projects were done on rivers, lakes and estuaries using one dimensional model and integrating it with remote sensing data, such as (Yang et al., 2000), where algal growth rate was modelled with QUAL2E, and 2D spatial data set derived from a SPOT image was used for calibration. In the present research one of the main objectives is to define the most appropriate way of calibrating water quality models with remote sensing.

8.3.2. Calibration Methods and Procedures

Both spatial and temporal calibrations were explored. The calibration methodology applying remote sensing included two levels. The first level is a qualitative calibration of the model results using patterns from the MODIS satellite image time series. A quantitative comparative analysis between the model-simulated patterns and the patterns resulting from MODIS time series analysis was conducted to check the model consistency with satellite data. This is an initial step in the remote sensing calibration application that is seen as a suitable indicator to judge the model performance on both spatial and temporal levels.

The second level of calibration is the quantitative calibration of the model. On this level of calibration, both datasets MODIS time series and the SPOT5 scene are used to develop TSM and CHL-a concentrations maps. Both data sets are used to compare the accuracy of the resulting concentration maps and subsequently the results of the model after application of these datasets. The procedures for calibrating the model are based on two selections: single point calibration (or specific target) and the whole map calibration (or the overall target). In the following sections the details of the different calibration levels, methodologies and procedures are explained along with the results of the model calibration and discussions on each method. Figure (8-1) shows the flow of the model adjustments and calibration processes.

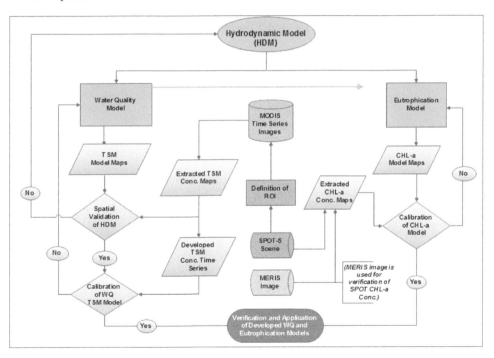

Figure (8-1): Flowchart of calibration processes using remote sensing data

8.4. SPATIAL PATTERN CALIBRATION FOR TSM USING MODIS IMAGES

The qualitative calibration is based on a comparison of the developed concentration patterns from the Delft3D water quality model and the patterns developed from the MODIS time series images during the modelling period of six months. The MODIS products MOD09GQK/Terra and MYD09GQK/Aqua are used. The specific characteristics of a MODIS image such as its medium spatial resolution, its red band reflectance, and its daily temporal coverage indicate that it may be well suited to examining suspended particulates (Miller and McKee, 2004). Proper sensor calibration, an accurate atmospheric correction and the removal of bottom interference are the three major challenges in moving towards an operational application of MODIS in water monitoring (Hu et al., 2004). Appendix (A-1) shows the main characteristics of the MODIS medium resolution bands.

8.4.1. MODIS Images Processing

A set of 18 MODIS images were selected to cover the modelling period of six months. The selection was limited to this number due to the intensive cloud coverage over the Delta region. A few cloud free images per month were suitable for use.

The data set of MODIS images was processed using the ENVI 4.3 software package (Environment for Visualization of Images). First, each image was corrected geometrically and geo-referenced to the UTM projection, Datum WGS-84 zone 36 N. MODIS images are quite large: an image covers almost half the area of Egypt. Therefore it was also necessary to create subsets to outline and resize the scenes of the region of interest from the raw images before applying other calculations and detailed analysis of the images as shown in Figure (8-2). Then, the images were screened and ranked in terms of quality, taking into account cloud coverage over the study area and other image errors. The selected images of the MOD09GQK series are level-2 products. Therefore no atmospheric correction was needed.

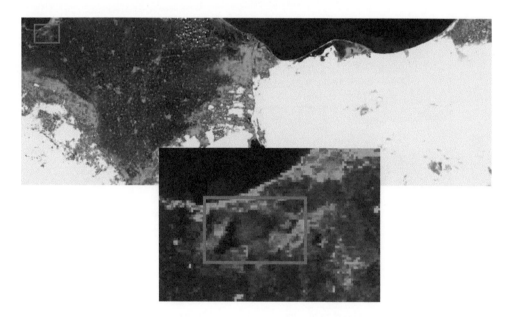

Figure (8-2): MOD09GQK 10th February, 2006 (part of full image and cropped image)

For the initial patterns comparison the model output grid maps are compared with the MODIS images having the same temporal properties i.e. the same dates and acquisition time of the images were selected from the model output maps. For the comparison, the model output maps were converted from the original Arc-GIS format (shape files) to raster maps with the same geometry as of the MODIS images. Both the MODIS images and the model raster maps were masked to the lake water body-Figure (8-3). A set of 12 selected images were used in the qualitative comparison-Figure (8-4).

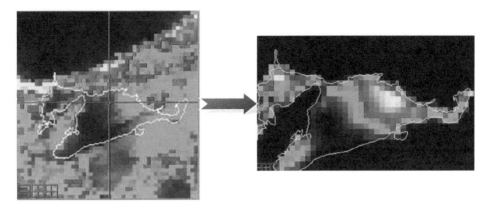

Figure (8-3): Resized MODIS image March18[th] (to the left) and masked image (to the right)

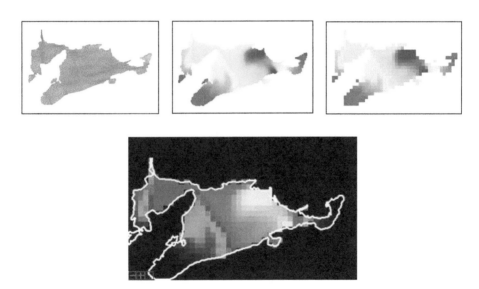

Figure (8-4): Processing steps of the Model output maps (from shape files to raster grid of 231.7 m resolution), (Example March 18th output map at 09:00 am)

8.4.2. Initial Qualitative Comparison of Profiles

The next step after processing each MODIS image and simulating the water quality parameters for the corresponding dates is the analytical comparison of each image pattern and its corresponding model pattern. McKee et al. (2004), developed a linear relationship between band 1 (620 – 670 nm) of MODIS Terra and in situ measurements of TSM, providing evidence of the transport and fate of material in coastal environments. Therefore, based on this linear relation the MODIS band 1 image was selected for the initial comparison of image and model patterns. Two sections were selected to develop profiles for comparison. The first section is from north-east to south-west direction, and the other is from south east to North West direction. Figure (8-5) shows the profiles for comparing the image and the model. The two images are staked using the model raster image and band one (B1) of the MODIS image, and a common transact is drawn on the staked image layers. The resulting profiles are shown in the following section, which includes a comparison of different patterns for some selected images as an example. Figures (8-6) to (8-11) show the spatial comparison of different patterns and the spatial profiles comparison for the days March18[th], April 11th and June 28th. Comparison between at surface reflectance from raw images with TSM Model concentration images, was done because the main target at this stage of calibration is to compare the trends of patterns from both data sets.

Figure (8-5): The selected sections for comparing the image and model spatial patterns

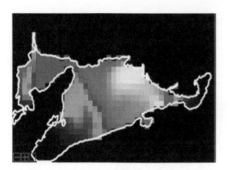 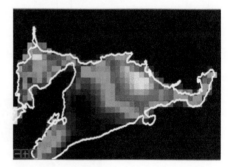

Figure (8-6): Model pattern (left) and image pattern (right), approx. 250m pixel size, March 18th 2006

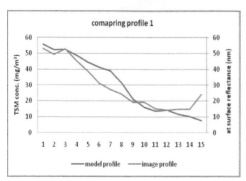 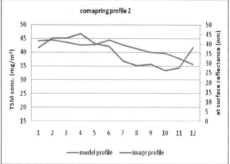

Figure (8-7): Profile1 (left) and profile2 (right), comparisons for the image acquired on March 18th and model raster output map

Figure (8-8): Model pattern (left) and image pattern (right), approx. 250m pixel size, April 11th 2006

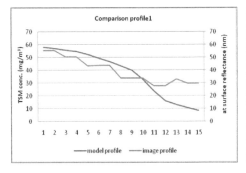 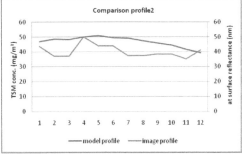

Figure (8-9): Profile1 (left) and profile2 (right), comparisons for the image acquired on April 11th and model raster output map

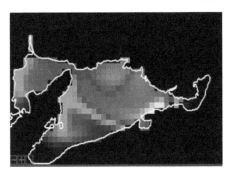 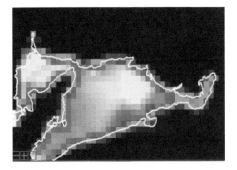

Figure (8-10): Model pattern (left) and image pattern (right), approx. 250m pixel size, June 28th 2006

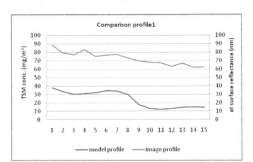 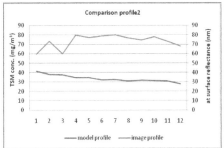

Figure (8-11): Profile1 (left) and profile2 (right), comparisons for the image acquired on June 28th and model raster output map

From the comparison of the profiles for the image and the model for the two selected sections, the above examples show in the most of the cases an acceptable trend of the model in comparison with the image profiles for the three selected dates. This level of qualitative calibration indicates that the model is predicting acceptable spatial profiles for the TSM concentrations, which can be observed as rough plume patterns in the images. Analysing the spatial patterns from the 14 selected images it is clear that there is a spatial and temporal variability of the patterns from one location to the other. Specific variations are obvious at the entrance points of the main drains to the lake and also at the far south western area. The 14 images representing seasonal variations of the flow patterns in the lake were staked in ascending temporal sequence (January-July) using the image processing programme ENVI, and a vertical common profile was drawn to represent the temporal variation at certain selected pixel locations.

Pixel to Pixel Comparison

A group of five locations was selected inside the lake, representing some of the measuring points of water quality samples namely (LO5, LO17, LO18, LO19, LO20), in order to test the variation. An averaging window is selected around each location. These averaged values were compared to the model data values at the same locations. The following section shows the compared averaged window pixels and the corresponding model values for the five selected locations. Figure (8-12) shows the selected measuring locations and the selected averaging pixels window.

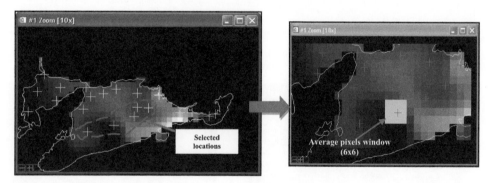

Figure (8-12): Selected measuring locations for temporal comparison and average pixels window (6x6) applied for each selected location

This qualitative calibration is based on a comparison of the time series of the model and the image based on original pixel values of the image before applying any analytical algorithm to the images. Figure (8-13-a,b,c,d) shows the compared image pixel values and the corresponding model pixel values; the trends are similar, e.g. LO5 and LO20, but there are variations within an acceptable range. The reasons for the variations can be found mainly in the uncertainty in georeferencing coarse resolution images that affects most of mixed pixels, the presence and dynamics of submerged and floating vegetation, and the uncertainties inherent in the atmospheric corrections of the MODIS L2 products.

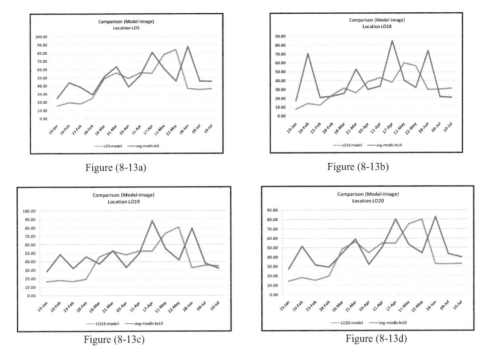

Figure (8-13 a,b,c,d): qualitative comparison of temporal images and model trends at different measuring locations inside the lake

Discussion on Results

The qualitative analysis using the preliminary comparison of images and modeled profiles can give a good indication of the general model trends and can be used as a first tool for judging the model performance. Also, comparing temporal trends using time series (pixel to pixel comparison) developed from the set of the model maps and images at different locations shows acceptable similar trends at some locations. The associated uncertainties with the atmospheric corrections and the presence of floating and submerged vegetation are seen as the main reasons for the variations at certain time steps between the image and the model pixel values. From the performed analysis we can conclude that qualitative calibration using MODIS could be used as an initial step before the detailed analysis of the images and before applying any algorithms to develop concentration maps.

8.5. SPATIAL AND TEMPORAL QUANTITATIVE CALIBRATION OF WATER QUALITY MODEL

At this level of calibration the images are used for quantitative analysis to extract water quality concentration maps for both TSM and CHL-a. The methodology is generally based on the application of an empirical or statistical approach for remote sensing data analysis. The maps are used to extract concentrations of TSM and CHL-a at both temporal and spatial levels to be used for model calibration. In this method of calibration the set of MODIS images is used to develop a time series data set for both TSM and CHL-a concentration maps that is then used as input data for calibrating the model at the main locations (namely, the exits of main drains into the lake water body).

8.5.1. Application of TSM Algorithms for development TSM Concentration maps

In literature several MODIS algorithms are developed for TSM extraction. These algorithms are site specific, and depend on the characteristics and the inherent optical properties (IOP's) of the water body. Previous work by Miller and McKee (2004) using the visible and NIR bands with the highest spatial resolution (250 m) provided useful results for total suspended matter (TSM) in the coastal waters off the northern Gulf of Mexico. The established linear relation was tested on the MODIS image of Lake Edko as shown in Equation (8-4):

$$\textbf{1140.25*(MODIS Band 1) – 1.91} \tag{8-4}$$

The extracted TSM concentrations applying this algorithm were very high compared to *in situ* measurements. The main reason for this difference is the different environments in the areas under analysis. Figure (8-14) shows the same previous selected pixel locations, the model predicted values, the image extracted values applying the Miller and McKee algorithm and the corresponding field measured TSM concentrations on 28th June. The comparison shows a difference between the measured values and the MODIS extracted values, with a RMSE of 33.27 mg/l. This shows that the developed algorithm is likely to be site specific and it is affected by the IOP's of the lake water.

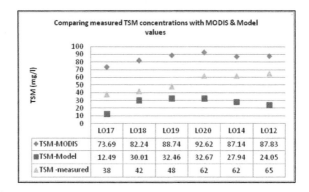

Figure (8-14): model predicted values, image extracted values and the corresponding field measured TSM concentrations on 28th June applying the algorithm developed by Miller and McKee, 2004

Applying the empirical approach to the analysis of remote sensing images as mentioned above is based on the calculation of a statistical relation between the water constituent concentrations and reflectance (or radiance). This approach needs extensive *in situ* data measurements of water quality parameter concentrations. To relate the number of measurements within the water body to the 250 m MODIS images in the case study under investigation, a limited number of measurements are available. Therefore a local regression analysis was done using the set of points that were measured on 28th of June 2006 when a MODIS image is available; for the rest of data set that were measured on 27th and 29th of June the corresponding images were with high cloud coverage and were excluded from the analysis. Since a similar case was developed for the turbid coastal waters of Mahakam Delta, Indonesia, (Budhiman, 2004), where it was proved that there is an exponential relation between the TSM concentration and MODIS band 1 (665 nm), the result of an exponential relation in the case of Edko Lake is considered to be accepted with a value of $R^2 = 0.85$ under the existing conditions. Figure (8-15) shows the regression analysis and the relationship between the measured TSM concentration and the surface reflectance at band 1 (665nm).

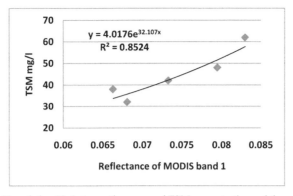

Figure (8-15): The relationship between the measured TSM concentration and the surface reflectance at band1 (665nm)

8.5.2. Selection of Region of Interest (ROI) Using SPOT High Resolution Scene

Due to the low resolution of the MODIS images which results in mixed pixels of water and vegetation even in open water areas, there was a need to define a specific region of interest (ROI) within the lake water body assuming the least possible mixing effect. This was done using the SPOT image with a resolution of 10m, where the region of interest was selected based on the free water pixels, and this area was up-scaled to the MODIS image. Figure (8-16) shows the original SPOT image of resolution 10m showing the detailed parts of the water body. The SPOT image and the defined regions of interest (ROI's) are shown in Figure (8-17). This selected ROI was used in all the comparisons of the detailed TSM model results and the image-based concentration maps.

Figure (8-16): SPOT image of resolution 10 meters showing detailed components of the lake system

Figure (8-17): SPOT image and selected regions of interest (ROI) within the main lake water body

8.5.3. Spatial Validation of Hydrodynamic Model Using TSM Patterns Comparison

As presented in the hydrodynamic modelling section, the model was calibrated for the Manning roughness coefficient. Since extra data sets for validation were not available, the remote sensing data was used in the validation of the model. Therefore, an initial step before spatial and temporal calibration procedures of the TSM and CHL-a models was to use the remote sensing data in the spatial validation of the hydrodynamic model. This was done through correlating the model time series TSM patterns and the MODIS time series based TSM maps. A correlation matrix was developed between both data sets and the descriptive statistical analysis results showed an average correlation coefficient value of $R^2 = 0.73$. Table (8-2) shows the developed correlation matrix. A correlation image was developed from the statistical analysis of the staked time series images that shows the spatial distribution of the correlation coefficients; Figure (8-18). The correlation image is a representation of the matrix. The dimension of the image is nb by nb where nb is the number of bands correlated or the number of bands of the input dataset.

Table (8-2): Correlation matrix of the model time series patterns and the MODIS time series

	B1	B2	B3	B4	B5	B6	B7	B8	B9	B10	B11	B12	B13	B14	B15	B16	B17	B18	B19	B20	B21	B22	B23	B24	B25	B26
B1	1.00	0.98	0.93	0.92	0.90	0.88	0.94	0.92	0.93	0.90	0.89	0.90	0.91	0.74	0.43	0.63	0.77	0.41	0.71	0.19	0.83	0.50	0.39	0.72	0.13	0.62
B2	0.98	1.00	0.94	0.92	0.90	0.88	0.94	0.92	0.93	0.91	0.89	0.90	0.91	0.74	0.42	0.63	0.75	0.41	0.69	0.19	0.83	0.49	0.37	0.71	0.11	0.59
B3	0.93	0.94	1.00	0.94	0.89	0.92	0.93	0.93	0.94	0.93	0.94	0.95	0.96	0.74	0.46	0.69	0.74	0.43	0.73	0.20	0.84	0.52	0.43	0.72	0.18	0.63
B4	0.92	0.92	0.94	1.00	0.98	0.96	0.98	0.99	0.99	0.99	0.97	0.97	0.96	0.74	0.58	0.72	0.83	0.55	0.77	0.26	0.83	0.61	0.52	0.80	0.26	0.73
B5	0.90	0.90	0.89	0.98	1.00	0.93	0.97	0.98	0.97	0.98	0.94	0.94	0.92	0.73	0.58	0.69	0.82	0.55	0.75	0.26	0.81	0.61	0.52	0.79	0.26	0.73
B6	0.88	0.88	0.92	0.96	0.93	1.00	0.96	0.97	0.97	0.97	0.97	0.99	0.99	0.76	0.64	0.80	0.82	0.59	0.80	0.30	0.87	0.66	0.55	0.83	0.28	0.75
B7	0.94	0.94	0.93	0.98	0.97	0.96	1.00	0.99	1.00	0.98	0.97	0.97	0.96	0.80	0.58	0.73	0.82	0.56	0.79	0.27	0.88	0.62	0.51	0.81	0.24	0.72
B8	0.92	0.92	0.93	0.99	0.98	0.97	0.99	1.00	1.00	0.99	0.98	0.98	0.97	0.77	0.61	0.75	0.83	0.58	0.79	0.28	0.86	0.64	0.54	0.82	0.26	0.74
B9	0.93	0.93	0.94	0.99	0.97	0.97	1.00	1.00	1.00	0.99	0.98	0.98	0.97	0.79	0.60	0.75	0.83	0.57	0.80	0.28	0.87	0.63	0.53	0.82	0.26	0.74
B10	0.90	0.91	0.93	0.99	0.98	0.97	0.98	0.99	0.99	1.00	0.98	0.98	0.97	0.76	0.61	0.76	0.83	0.58	0.79	0.29	0.85	0.65	0.55	0.82	0.28	0.75
B11	0.89	0.89	0.94	0.97	0.94	0.99	0.97	0.98	0.98	0.98	1.00	1.00	0.98	0.76	0.63	0.79	0.82	0.59	0.81	0.29	0.87	0.66	0.56	0.83	0.29	0.75
B12	0.90	0.90	0.95	0.97	0.94	0.99	0.97	0.98	0.98	0.98	1.00	1.00	0.99	0.77	0.62	0.79	0.82	0.57	0.80	0.28	0.87	0.65	0.54	0.82	0.27	0.74
B13	0.91	0.91	0.96	0.96	0.92	0.98	0.96	0.97	0.97	0.97	0.98	0.99	1.00	0.75	0.58	0.77	0.79	0.53	0.77	0.27	0.86	0.61	0.51	0.79	0.23	0.71
B14	0.74	0.74	0.74	0.74	0.73	0.76	0.80	0.77	0.79	0.76	0.76	0.77	0.75	1.00	0.61	0.78	0.76	0.63	0.88	0.33	0.93	0.65	0.62	0.79	0.33	0.74
B15	0.43	0.42	0.46	0.58	0.58	0.64	0.58	0.61	0.60	0.61	0.63	0.62	0.58	0.61	1.00	0.81	0.76	0.89	0.80	0.59	0.69	0.88	0.73	0.85	0.45	0.77
B16	0.63	0.63	0.69	0.72	0.69	0.80	0.73	0.75	0.75	0.76	0.79	0.79	0.77	0.78	0.81	1.00	0.87	0.80	0.89	0.57	0.84	0.86	0.79	0.90	0.52	0.89
B17	0.77	0.75	0.74	0.83	0.82	0.82	0.82	0.83	0.83	0.83	0.82	0.82	0.79	0.76	0.76	0.87	1.00	0.74	0.91	0.47	0.84	0.85	0.77	0.92	0.44	0.91
B18	0.41	0.41	0.43	0.55	0.55	0.59	0.56	0.58	0.57	0.58	0.59	0.57	0.53	0.63	0.89	0.80	0.74	1.00	0.79	0.65	0.68	0.89	0.74	0.84	0.52	0.76
B19	0.71	0.69	0.73	0.77	0.75	0.80	0.79	0.79	0.80	0.79	0.81	0.80	0.77	0.88	0.80	0.89	0.91	0.79	1.00	0.54	0.90	0.83	0.79	0.89	0.51	0.88
B20	0.19	0.19	0.20	0.26	0.26	0.30	0.27	0.28	0.28	0.29	0.29	0.28	0.27	0.33	0.59	0.57	0.47	0.65	0.54	1.00	0.34	0.61	0.45	0.47	0.65	0.50
B21	0.83	0.83	0.84	0.83	0.81	0.87	0.88	0.86	0.87	0.85	0.87	0.87	0.86	0.93	0.69	0.84	0.84	0.68	0.90	0.34	1.00	0.72	0.63	0.88	0.29	0.78
B22	0.50	0.49	0.52	0.61	0.61	0.66	0.62	0.64	0.63	0.65	0.66	0.65	0.61	0.65	0.88	0.86	0.85	0.89	0.83	0.61	0.72	1.00	0.82	0.91	0.54	0.82
B23	0.39	0.37	0.43	0.52	0.52	0.55	0.51	0.54	0.53	0.55	0.56	0.54	0.51	0.62	0.73	0.79	0.77	0.74	0.79	0.45	0.63	0.82	1.00	0.73	0.47	0.83
B24	0.72	0.71	0.72	0.80	0.79	0.83	0.81	0.82	0.82	0.82	0.83	0.82	0.79	0.79	0.85	0.90	0.92	0.84	0.89	0.47	0.88	0.91	0.73	1.00	0.42	0.88
B25	0.13	0.11	0.18	0.26	0.26	0.28	0.24	0.26	0.26	0.28	0.29	0.27	0.23	0.33	0.45	0.52	0.44	0.52	0.51	0.65	0.29	0.54	0.47	0.42	1.00	0.66
B26	0.62	0.59	0.63	0.73	0.73	0.75	0.72	0.74	0.74	0.75	0.75	0.74	0.71	0.74	0.77	0.89	0.91	0.76	0.88	0.50	0.78	0.82	0.83	0.88	0.66	1.00

0
0.1
0.2
0.3
0.4
0.6
0.7
0.8
0.9
1

Figure (8-18): The correlation image of the whole dataset including
Model result and MODIS time series

The spatial distribution of the TSM patterns is highly affected by the hydrodynamic characteristics of the model, which is mainly forced by wind as described in the hydrodynamic modelling section. In the hydrodynamic model the calibration target was water levels at specific measured points. The comparison of the TSM patterns between the model and the image was used to set the validation target parameter from the remote sensing images (MODIS time series) based on the sensitivity of TSM spatial distribution to velocity patterns. The average correlation coefficient of 0.73 is considered valid for the spatial comparison between the two datasets taking into account the effect of mixed pixels and calibration accuracy of the received MODIS raw images. This spatial correlation indicates that the TSM patterns of the model and the images are positively correlated and follow the same trend. Consequently we can consider the hydrodynamic model to be a good performing model that can be used in the next steps of calibration and validation.

8.5.4. Developing TSM Time Series dataset from TSM-MODIS concentration maps

The developed correlation formula shown in Equation (8-5) was tested on the sampling points of open water areas and applied on the set of 14 images covering the modelling period. Focused analysis was done on the image acquired on June 28th where the field measurements were taken, to compare the values from the *in-situ* measurements, the image and the model. The image reflectance values were converted to suspended sediment concentration values. Figure (8-19) shows TSM map using the developed exponential relation.

$$CHL\text{-}a = 4.0176 \, EXP \, (32.107B1) \qquad\qquad (8\text{-}5)$$

Where:
CHL-a = chlorophyll a concentration (mg/l)
B1= MODIS Band 1 (665 nm)

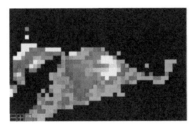

Figure (8-19): MODIS TSM concentration map June 28th applying the developed algorithm

In order to compare the model and image patterns, the ROI was applied to both maps, Figure (8-20 a and b). The two maps were stacked for correlation analysis. The resulting correlation coefficient of 0.89 shows high spatial similarity. On the other hand, comparing of the two profiles taken from both maps shows that the model is underestimating the concentration values, Figure (8-21). Selected locations for pixel to pixel comparison are shown in Figure (8-22) including the model predicted TSM concentrations on 28th of June, 2006.

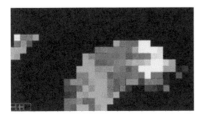 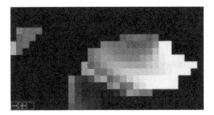

Figure (8-20): (a) TSM concentration map applying the developed algorithm (left) and (b) TSM concentration map predicted from model

177

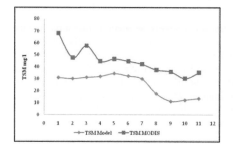 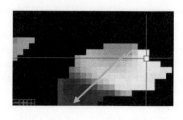

Figure (8-21): profile comparison of developed TSM map and the TSM model predicted values

To test the validation of the developed algorithm, a pixel to pixel comparison was done between the model and the satellite-based TSM maps; at selected *in situ* measurement locations. These samples locations were selected for analysis because they are measured on the same date as the available image for month June (28th June). The other measured points were taken on 27th and 29th June, but unfortunately the MODIS images acquired on the same dates were covered with intensive cloud and were not suitable for analysis. Two points (LO12, LO14) were used for the validation of the developed relation that were not included in the calculation of regression relation. Figure (8-22) shows the selected pixel locations.

Figure (8-22) Selected locations for pixel-to-pixel comparison

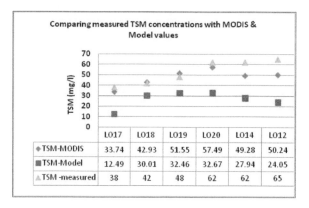

	LO17	LO18	LO19	LO20	LO14	LO12
◆TSM-MODIS	33.74	42.93	51.55	57.49	49.28	50.24
■TSM-Model	12.49	30.01	32.46	32.67	27.94	24.05
▲TSM -measured	38	42	48	62	62	65

Figure (8-23): Model predicted values, image extracted values and the corresponding field measured TSM concentrations on 28th June

Figure (8-23) illustrates the relationship between the three data sets; a positive trend is observed. It is concluded that the model is underestimating the measured values and the MODIS extracted values. Calculating the RMSE between the measured and MODIS TSM values is 2 mg/l, and between the measured and modelled TSM values is 5.12 mg/l. The developed TSM relation could be used to develop a TSM time series by applying it to the set of images covering the modelling period. It is noted that the model is under predicting the TSM concentrations compared to the measured and extracted values, but it is still considered within an acceptable range of RMSE. Therefore the developed time series could be used to calibrate the model through intensifying the model forcing data. The selected locations for developing the TSM time series are the exits of Edko and Barseek drains into the lake. To develop the time series from the images, first the algorithm is applied to all the images, and then the layer stacking tool under ENVI is used to form a stack of layered images representing the temporal sequence of the images. A vertical spectral profile is drawn intersecting these grouped images at the selected pixel locations to represent the time series. Figures (8-24) shows the developed time series at both locations.

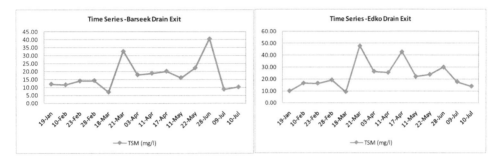

Figure (8-24): TSM concentrations time series from MODIS TSM maps at Edko and Barseek drains exits into the lake

8.5.5. Calibration of Model Using Developed MODIS TSM Concentration Maps

The mathematical model was calibrated for TSM using the developed time series as a new input dataset at the exit points of both drains entering the lake as shown in Figure (8-25). The new predicted values are changed slightly are due to the effect of the accuracy of the developed algorithm and the site characteristics which need detailed information on inherent optical properties (IOP's) of the water body under investigation. Also some spectral measurements taken during water quality sampling reflect the mixed properties of the water body since the shallow lake is dominated by the presence of the submerged vegetation Potamogeton pectinatus, and this can affect the measured concentrations of TSM.

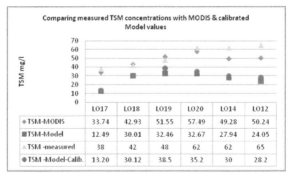

	LO17	LO18	LO19	LO20	LO14	LO12
◆TSM-MODIS	33.74	42.93	51.55	57.49	49.28	50.24
■TSM-Model	12.49	30.01	32.46	32.67	27.94	24.05
▲TSM -measured	38	42	48	62	62	65
●TSM -Model-Calib.	13.20	30.12	38.5	35.2	30	28.2

Figure (8-25): Model calibrated values, image extracted values and the corresponding field measured TSM concentrations on 28th June

Although the quantitative calibration of TSM did not show a big variation, the same trend is well recognized between the model and the image values. The important result from the calibration using the image data is that it showed close relation to the field measurements and it showed that the mathematical model could be insensitive to the TSM re-suspension. This explains the underestimated concentrations from the model. More detailed analysis could be done using analytical remote sensing approach or bio-optical modelling to develop more accurately calibrated algorithms for the TSM maps because the remote sensing of TSM involves a big level of uncertainty that may occur due to the presence of intensive submerged and floating vegetation in the lake and this may reflect higher measured signals of TSM. Further investigation is needed to develop an accurate algorithm to estimate suspended sediment concentration over Lake Edko using MODIS images. In order to achieve accurate estimates it is necessary to develop a site-specific algorithm that fits the conditions of the Lake.

8.6. SPATIAL AND TEMPORAL QUANTITATIVE CALIBRATION OF THE SCREENING EUTROPHICATION MODEL

8.6.1. Chlorophyll-a Retrieval from Remotely Sensed Imagery

CHL-a is not uniformly distributed in inland waters. This light-absorbing molecule in the chloroplasts of algal cells is the universal algal pigment, and CHL-a is a simple measure of phytoplankton biomass in surface waters. Thus, as mentioned above, we use CHL-a as an indicator of eutrophication in this research.

Remote sensing of CHL-a has had limited success in turbid productive coastal waters. Far offshore, in *case I waters*, ocean color (upwelling radiance) is largely determined by water and CHL-a, and extraction of CHL-a from ocean color measurements has been done successfully (e.g., Gordon and Morel 1983). However, in turbid productive coastal waters known as *case II waters*, color is determined by water, CHL-a, colored dissolved organic matter (CDOM), and non-phytoplanktonic particulates (seston or tripton). Although inland waters (e.g., lakes, rivers, reservoirs) usually have a higher range of CHL-a and thus a stronger signal, the independent variations in CDOM and particulates have impeded the routine extraction of CHL-a from ocean color measurements inshore. We refer to turbid waters with high CHL-a concentrations as *"turbid, productive waters"* whether inland or at the land margins. Most inland waters are classified as Case 2 waters (Morel and Prieur 1977; Kirk 1994). Much of the research on Case 2 waters has been focused on methods for analyzing the reflectance as a function of the concentrations of phytoplankton groups, inorganic matter, and dissolved organic matter. Most of the algorithms developed for estimating CHL-aorophyll concentration from inland aquatic systems are based on reflectance spectra derived from ground measurements and airborne sensors (Lathrop and Lillesand 1986; Dekker 1993; Novo et al. 1991) or from satellite sensors not tuned to the radiometric and spectral resolution requirements of Case 2 waters. The algorithms are generally site specific and empirically derived through statistical relationships between reflectance and CHL-a concentration.

Although the MODIS images used in this research have (250m) resolution they have only two bands; one red and one near –infrared. The lack of green band makes this MODIS data set more suitable for the TSM analysis, but it is not applicable for the CHL-a extraction. Therefore, the SPOT-5 imagery is adopted to derive the calibration data, because of its high spatial resolution (10 m) and useful spectral resolution (including the green band). The SPOT scene was used as guidance for concentrations of CHL-a at that time of the year since there was no available scene at the same time of the field measurements. The image was taken twelve days after the field work measurements, on the 10th of July 2006, but it was still comparable to the modelling results and the data available from literature. To have a practically applicable SPOT scene for water quality variables extraction the image has to pass several image processing steps. The following section explains the detailed SPOT image processing steps to convert digital values (DN) of the SPOT satellite images to water quality variables. Figure (8-26) shows the raw SPOT scene.

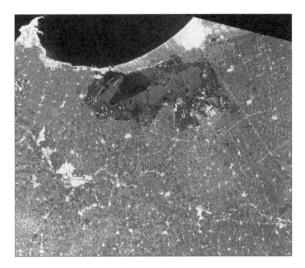

Figure (8-26): SPOT-5 full scene.

SPOT images are delivered in digital numbers format, and the initial step in processing these images is to convert the digital numbers into radiance, then to convert the radiance into at surface reflectance. The first conversion is simple linear transformation using sensor-dependent parameters, and the second conversion was done for atmospheric correction applying radiative transfer algorithms. The atmospheric correction of the image was done using the ENVI's Fast Line-of-sight Atmospheric Analysis of Spectral Hypercubes (FLAASH) module. The detailed procedures for atmospheric correction are explained in Appendix (A-2). The atmospheric correction is an important image calibration process to remove the effects of aerosols and water vapour effects. The atmospherically corrected image was then resized to the main lake area; Figure (8-27) shows the atmospherically corrected resized image of the lake area.

Figure (8-27): The atmospherically corrected resized SPOT image of the lake area.

The next step in the image processing was masking out of the open water body from the image taking separating it from the mixed pixels including water, floating and submerged vegetation in the lake. This was done using unsupervised classification. In this way the water pixels were differentiated from the other pixels. An unsupervised classification was applied using the SPOT band 3 (NIR) image

181

according to Yang et al 2000. This was done by assigning the first class to water throughout the entire image. The procedure developed binary mask water pixels as "1" and the rest of the pixels as "0". Afterward, the water body was extracted from the other SPOT bands as shown in Figure (8-28).

Figure (8-28): Masked water body SPOT image

8.6.2. Application of CHL -a Algorithms to develop CHL-a Concentration maps

To develop the CHL-a concentration maps, a natural logarithmic band ratio regression model developed by (Yang et al 2000) was applied to the SPOT image. This band ratio is between the near infrared (NIR – X3) and (red-X2) bands due to a positive reflectivity of CHL-aorophyll in the NIR and an inverse behavior in the red (Rundquist et al., 1996). This regression model used is as shown in equation (8-6).

$$Ln\ CHL\text{-}a = 9.37 + 10.10\ ln\ XS3/XS2 \qquad (8\text{-}6)$$

This regression model was applied on the SPOT image scene of the lake. The resulting concentrations map is shown in Figure (8-29).

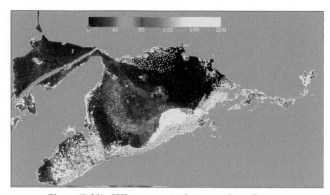

Figure (8-29): CHL-a concentrations map (in µg/l)

The results from the image processing and application of the selected algorithm in comparison with the field investigations shows that there are areas of mixed pixels of water and vegetation due to intensive amounts of floating and submerged vegetation in the water body, which accounts for high values of CHL-a that exceed 100 µg/l specially in the middle area of the lake and around the lake shores where floating mats of vegetation are manually constructed by fishermen as traps for fish. This is despite the fact that the average expected values at this time of the year (during the summer season the average concentration is 58 µg/l in the mid area of the lake, and 7.58 µg/l in the western region of the lake, according to (Siam et al, 2000) and the values deduced from field work conducted on May 2005 showed an increase in the average concentration to 61 µg/l in the mid of the lake. It is obvious from the field observations that during this period of the year there is an effect by the floating and submerged vegetation on the CHL-a concentrations retrieved from the image.

Comparing the readings from the image at specific locations where it was possible to take field measurements (LO11, LO12, LO13, LO14, LO15, and LO21) where the readings range from 31.14 – 86.7 µg/l to the limited measured values (11.3-24.4 µg/l) it is deduced that the concentrations from the image were reasonably higher. Figure (8-30) shows the locations where CHL-a samples were measured. In recent publication by (Ossman et al, 2010), concentrations of CHL-a within lake Edko recorded higher values that ranges between (28-99 µg/l). Taking into account the limited number of measured values in 2006 and the values from earlier field work in 2005 and published CHL-a concentrations in Lake Edko, it is concluded that the CHL-a in open water is in the range of 40-76 µg/l in the middle region of the lake. The following uncertainty sources should be highlighted; the image taken was 11 days later than the measurements values, no samples were taken in the middle area of the lake, submerged and floating vegetation may result in higher CHL-a readings.

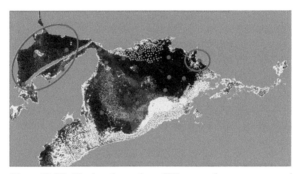

Figure (8-30): The locations where CHL-a samples were measured

The average concentration of CHL-a within the (ROI) is 31.4 µg/l within the whole (ROI) in both the easernt and western areas of lake water body this is shown in Figure (8-31).

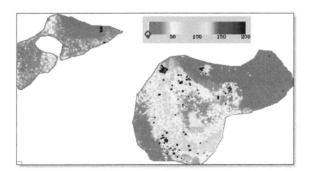

Figure (8-31): Concentrations of CHL-a within the selected (ROI)

It can be concluded from the analysis of the SPOT image, that the CHL-a concentration in the open water area of the lake was in the same range as it is expected from the data published in the literature. The high spatial resolution of the SPOT image proved that there are small islands of floating vegetation and submerged vegetation areas; Figure (8-31), which will give increased estimated Chl-values in the images of coarser resolutions. This single SPOT image provided a good insight into the state of chlorophyll distributions on the 10th of July, 2006. Before using the extracted SPOT concentrations as a reference for the model mathematical model calibration, in the next step, the SPOT-based chlorophyll values were upscaled to the MERIS image (Medium Resolution Imaging Spectrometer) captured at the same exact date of the SPOT (10th of july, 2006). The main objective of this step was to compare and validate the used algorithm for CHL-a extraction from SPOT image.

8.6.3. Validation of SPOT-5 Extracted CHL-a Concentrations: Using (MERIS Satellite Image)

MERIS sensor is one of the European Space Agency (ESA) sensors onboard the Envisat environmental satellite. MERIS sensor measures the solar radiation reflected by the Earth in 15 spectral bands, ranging from the visible to the near-infrared spectral regions (390 nm to 1040 nm) at a maximum ground spatial resolution of 300 m, and provides the most radiometrically accurate data on Earth surface that is currently acquired from space (Curran and Steele 2005). MERIS data is known for its relatively high spatial resolution and spectral bands optimized for estimating water quality parameters. See Appendix (A-3) for MERIS images characteristics.

A full-resolution geo-located Level 1b MERIS FR image was acquired on July 10th 2006, the same date as the SPOT image described in 8.6.2. A subset from the image including Lake Edko area was selected for analysis Figure (8-32). The image was analyzed using BEAM-4.9 software (Brockmann Consult, Geesthacht, Germany). BEAM is an open-source toolbox and development platform for viewing, analysing and processing of remote sensing raster data such as MERIS. The BEAM software includes a plug-in algorithm the MERIS case 2 water processor (MERIS C2R), which is a global processor for retrieval of water quality parameters concentrations including TSM and CHL-a. This plug-in algorithm is developed by Doerffer and Shiller (2007) especially for analysis of Case 2 Coastal Waters and it is based on the inversion of the radiative transfer model using artificial neural networks, so it is using a physically based observation model. The algorithms relate the radiances observed by MERIS to first atmospherically corrected reflectances and then to water quality constituents.

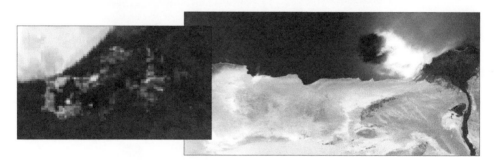

Figure (8-32): MERIS full resolution scene , July 10th 2006 and the subset scene of Lake Edko

The plug in processor MERIS C2R of the BEAM software is applied on the subset image scene. The processing includes several steps, the first step is the atmospheric correction of the image and the second one is the water algorithm calculating the TSM and the CHL-a concentrations. Figure (8-33) shows the resulting CHL-a concentration map from the MERIS image for all water bodies including the fisheries area. A masked subset of the CHL-a map is shown in Figure (8-34). The pixel values of the subset showed an average concentration 19 µg/l with a value range of 0.4-35µg/l. The eastern lake CHL-a concentrations are around 33µg/l. The comparison proves that the CHL-a map derived from the SPOT image with an empirical model is in good correspondence with the CHL-a map derived from the MERIS image with a physically based observation model. It can be concluded that the SPOT-based map is suitable for being used for the calibration of the CHL-a modelling parameters in the Delft-3D model.

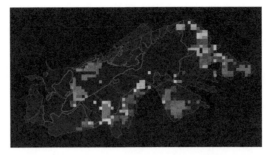

Figure (8-33): CHL-a map of the lake region (including vegetation and fisheries areas). For Colour scheme see Figure (8-34)

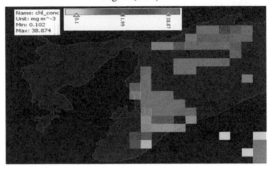

Figure (8-34): Extracted CHL-a pixels inside the lake water body

8.6.4. Calibration of CHL-a Model Using SPOT-Based Concentration Maps

The calibration of the CHL-a model was done based on adjusting the algal species to nutrients ratios. Due to the spatially limited number of measured CHL-a samples in the lake and the lack of any historical measurements of CHL-a there was a need to use data from the remote sensing images it the calibration process. The model calibration was done based on adjusting the nutrients inputs to develop a CHL-a map with concentrations close to the images extracted concentrations. Several runs were carried out by adjusting the nutrients inputs until the concentrations of CHL-a became close to the extracted values from the SPOT image. Figure (8-35) shows the CHL-a model results prior to calibration procedures.

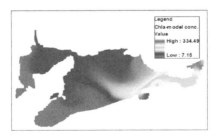

Figure (8-35): CHL-a model concentrations (July 10th, 2006

The first step in the calibration of the CHL-a model is to test the spatial variation between the model and the image maps. Two sections were taken along the flow paths shown in Figure (8-36) to illustrate the spatial variation in the concentrations along the northern area of the lake region free of submerged

185

or floating vegetation. The same profiles were taken from the CHL-a model map to compare the spatial variation and to highlight the spatial resolution difference between the model and the SPOT Data taking into consideration that the model map which is (50x50m) was resized to the same pixel size (10x10 m) of SPOT image.

Figure (8-36): The profiles paths taken on SPOT image

The spatial profiles comparison shows a high spatial variation of CHL-a concentrations from the SPOT images along the selected sections in the ranges shown in Table (8-3), this is a result of the high spatial resolution of the SPOT image. The model map shows a smooth variation along the same paths (with higher concentrations than extracted from the image as mentioned earlier as shown in Figure (8-37).

Table (8-3): CHL-a conc. ranges along selected profiles

Profile	Max	Min	AVG
SPOT - p1	111.72	3.86	33.73
Model-p1	108.45	72.15	88.36
SPOT - p2	99.13	15.36	38.29
Model-p2	76.71	57.80	69.83

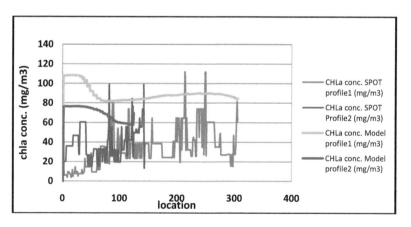

Figure (8-37): Comparison of CHL-a concentrations: Model and SPOT spatial profiles before model calibration

The calibration of the model was done based on adjusting the concentrations of the nutrients and related algae species concentrations inputs and comparing the results with the extracted and verified concentrations ranges from SPOT image. The nutrients involved in calibration process are the nitrates, ammonium and phosphorous compounds in addition to the diatoms and green algae ratios to these nutrients. The ratios of the algae species were reduced by a varying percentage to reach the remote sensing observed concentrations ranges at that period of the year. Reduced ratios of algae species to the nutrients with a range from 10% to 60% were tested and results from model were compared with the expected ranges from literature. Decreasing the ratios of algae species to the nutrients by 20% resulted in an average of 32.13 mg/l. These values are considered within the range of extracted SPOT concentrations. Figure (8-38) shows the model CHL-a concentrations after reducing nutrients–algae species ratios. The spatial distribution of CHL-a concentrations after model calibration are shown in Figure (8-39).

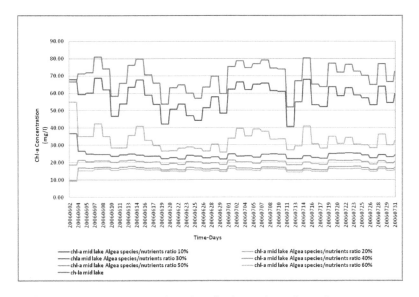

Figure (8-38): CHL-a concentrations after adjusting nutrients–algae ratio parameters

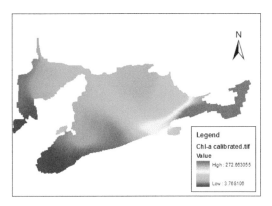

Figure (8-39): Calibrated CHL-a concentrations map, July, 10th, 2006

Due to the limited number of measured CHL-a, the calibrated model data is compared with the image data for the same sections (1,2) as shown in Figures (8-40) and (8-41), it is indicated that the model concentrations are reduced to reach the average concentrations of the SPOT image specially at the northern eastern part of the lake open water area.

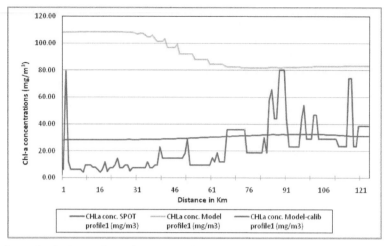

Figure (8-40): Comparison of CHL-a Model and SPOT concentrations (Profile 1)

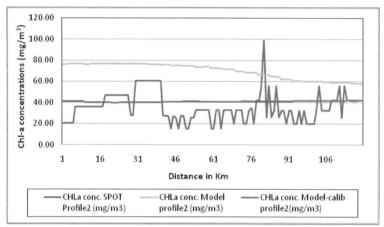

Figure (8-41): Comparison of CHL-a Model and SPOT concentrations (Profile 2)

Table (8-4) shows a qualitative evaluation of the calibrated model using the comparison of selected CHL-a SPOT image profiles (1,2) to the model values and the calibrated model values, the average values and the statistical analysis of mean average error (MAE) and root mean squared error (RMSE). The calculated errors are considered within an acceptable range taking into account the spatial resolution difference between the image and the model.

Table (8-4): statistical analysis between predicted calibrated CHL-a model and earth observation CHL-a derived from SPOT image

Profile	Max	Min	AVG	MAE	RMSE
SPOT - p1	111.72	3.86	33.73		
Model-p1	108.45	72.15	88.36	90.76	14.30
Model-calib-p1	32.46	28.02	30.20	44.68	6.70
SPOT - p2	99.13	15.36	38.29		
Model-p2	76.71	57.80	69.83	51.68	12.65
Model-calib-p2	42.15	39.67	41.02	13.98	3.88

8.7. UNCERTAINTIES WITHIN THE APPLIED TOOLS AND METHODS

Uncertainty analysis for model simulation is of growing importance in the field of water quality management. In water quality modelling, sources of uncertainty could be; error due to inaccuracies of the input data, model structure, uncertainty in the estimated model parameter values, and the propagation of prediction errors. Added to these main sources, the complexity of the system under study and the integration of different methodologies, modelling tools calibration and verification tools could be another added indirect sources of uncertainty. The integration of knowledge in the form of mathematical modelling, remote sensing modelling and earth observation data can be very useful for a variety of reasons: i) such models and tools make it possible to test hypotheses on functional interactions in the systems under investigation. (ii) they are tools for interpreting knowledge between different users for management and decision support. (iii) they can be used for predicting future states of the systems or its responses to assumed or expected changes in driving conditions. iv) they can be used for forecasting of future extreme events. In the mean time the complex the modelling system and tools gets the more uncertain predictions might be associated. In this research study all the mentioned sources of uncertainty are involved within the used and/or developed tools, research methods and data acquisition sources. The main driver of this research was to make use of integrating mathematical modelling, GIS and remote sensing tools for surface water quality management in *"data scarce"* irrigated watersheds. Data scarcity is the first uncertainty associated problem in water quality modelling where very limited datasets were available but were not enough to conduct detailed sensitivity analysis for all developed models. The used modelling tools are physically based models and these models involve large number of variables modelling parameters and coefficients in both hydrodynamic and water quality models which further contribute to uncertainty and error propagation. The complexity of the lake water system and the numerous interacting eutrophication model coefficients, variables, physical, chemical and biological processes are considered another important source of modelling uncertainty. Remote sensing was used for data acquisition and for extraction and prediction of water quality parameters that were used for calibration and verification of the mathematical models. Remote sensing procedures and methods itself involves uncertainty that includes temporal and spatial variability of input data, atmospheric correction accuracy, applied models and algorithms, accuracy of extracted data and ground truth verification data. Uncertainty analysis of the different modelling systems should be conducted for models fine-tuning and verification.

Lack of continuous temporal measurements, neglected measurements of important eutrophication indicators, lack of archived historical data of the lake system was a major constraint in the research and in the mean time it was one of the driving forces to conduct the research based on the integration of tools and data acquisition sources for calibration and verification of models. Although there is a level of error propagation in the different linked modelling tools specially the eutrophication model, the developed 2D hydrodynamic model, the basic water quality model and the screening eutrophication models can be considered robust and reliable for their designated management objectives since the calculated errors are within accepted ranges. Uncertainty analysis was not a in the scope objectives of this research but it is a priority for the use of the system future water quality and eutrohphication forecasting of the watershed lake system.

9. APPLICATION OF THE DEVELOPED (WQMIS) AS A DECISION SUPPORT TOOL FOR LAKE MANAGEMENT

The general objective of this research work was to develop a practically applicable Surface Water Quality Management Information System WQMIS (including the assessment, modelling and management components) that is applied at a watershed scale focusing on catchment –lake system components. The developed DS tools aims to answers both planning and technical questions of water quality managers, decision makers, and those of technical engineers working on the sampling, monitoring, analysis and modelling of water quality parameters. This system is developed as an integrated computational framework, and the different developed modelling tools are meant to be used for management purposes. This chapter focuses on the application of the developed water quality and eutrophication models in developing management scenarios for reduction of nutrients entering the lake system from different anthropogenic activities within the watershed specifically the agriculture activities.

9.1. DECISION SUPPORT FOR LAKE WATER MANAGEMENT

The developed WQMI includes a structured set of decision support tools for managing the watershed such as the surface water quality geo-database, the 1D and 2D hydrodynamic models of the catchment and lake system, the basic water quality modelling tool of the lake and the screening eutrophication model of the lake. In this research study a main target was to create better understanding of the complex water system and to provide useful and reliable knowledge for managing the watershed and its components. The focus of the research was on the surface water quality management and specifically the lake water system since it is the collector of all pollutants and wastes discharging from the catchment. The main objectives of the DS tools are: The DSS will be an important tool to: to increase the understanding of the relations between different users and the components of the water system to integrate the different tools to overcome the lack of data and to provide a common and user-friendly framework for the analysis and comparison of management options and measures though using a set of physically based models supported by remote sensing data for calibration and validation procedures. The developed set of water quality modelling tools is mainly focusing on prediction of water quality concentrations and eutrophication indicators concentrations to assist in management of the lake system. A plan for managing the water quality of the lake system can be implemented using these models.

9.2. MANAGEMENT PREDICTION SCENARIOS FOR SHALLOW LAKE SYSTEM

The main function of the developed water quality and eutrophication modelling tools is to produce predicted water quality concentrations for the assigned group of parameters specially the TSM and in order to develop the possible management plans for lake pollution control and mitigation strategy. The screening eutrophication model could be also used for predicting the CHL-a seasonal ranges for management practices. As mentioned, discussed and concluded from this study, the main pollutants entering the lake are from non point sources in the form of agricultural nutrients; mainly nitrogen and phosphorous compounds, in addition to untreated wastewater that raise the concentrations of TSM. In this research study a nutrient loads reduction methodology is presented, but for detailed and strategic planning a wide range of temporal and spatial data sets will be needed.

As seen from chapter 5 the analysis results of the nutrients loads entering the lake shows amounts of 1080 ton/year of TP and 35×10^3 ton/year of TN. To reduce the pollutants loads to the lake and to prevent more deterioration of lake water quality a set of planned scenarios for nutrients reduction is developed and tested. The two main drains contributing to the lake pollution are Edko and Barseek drains so the reduction of nutrients coming from these sources was investigated for pollution control. The first set of developed scenarios involves equal reduction factors of nutrient loads coming from both drains. The second set of scenarios involved only the reduction of pollutants coming from Barseek drain which contributes by higher amounts of nutrients and wastes to the lake water, (reduction factors from 0.05 to 0.5 were used in these scenarios).

It is also known that the aquatic vegetation biomass increases by the uptake of available phosphorus and nitrogen from the water. It was found that the nutrient that will control the maximum amount of plant biomass is the nutrient that reaches a minimum before other nutrients. Therefore, under certain condition, nitrogen may reach a minimum value before phosphorus and, as a result, control the maximum amount of plant biomass and vice versa. This situation depends on the relative amounts of nitrogen and phosphorus required by aquatic plants and their availability in the water body. Accordingly, a mass ratio of available forms of nitrogen and phosphorus (N/P) was used to calculate the limiting nutrient in water. But from the modelling results this information could be also retrieved based on the reduction scenarios that will be presented discussed below.

A. Reduction of Nutrients Loads from all Drainage Points to the Lake

Since one of the main objectives of this study is to understand the effect of the pollutants loads discharging from the watershed to the lake, the focus of the prediction scenarios was on the lake system and how to reduce the nutrients loads to the lake which will imply strategic planning actions to be taken at the upstream of the watershed. At the boundary entrances of the lake (Edko and Barseek drains outlets) different reduction percentages of nitrates NO3, ammonia NH4 and phosphorus PO4 were applied on the eutrophication model. Reduction percentages of, 10%, 30% and 50% were used for the three main nutrients at the two location of Edko and Barseek. Figure (9-1) shows the effect of reduction of all nutrients on the CHL-a concentrations a reduction of 10% of the nutrients (NO3, NH4 and PO4) gives a concentration average around 20 mg/m^3, reduction of 30% and 50% gives almost a constant temporal variation of CHL-a over the modelling time. These reductions should be further tested at different locations for verification.

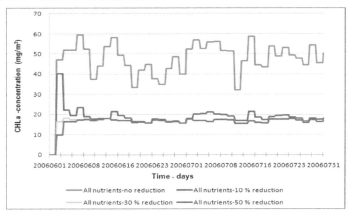

Figure (9-1): The effect of reduction of nutrients on the CHL-a concentrations at mid lake location

B. Reduction of phosphorous Loads from all Drainage Points to the Lake

Reducing only the phosphorous nutrients by the same percentages used shows a similar reduction in the CHL-a concentrations in comparison to the reduction of all nutrients components, Figure (9-2). A reduction of 10% of the phosphorous (PO4) gives a concentration average around 20 mg/m^3, reduction of 30% and 50% gives almost a constant temporal variation of CHL-a over the modelling time.

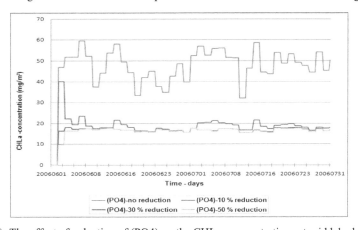

Figure (9-2): The effect of reduction of (PO4) on the CHL-a concentrations at mid lake location

193

C. Reduction of phosphorous Loads from Barseek drain to the Lake

Since most of the fisheries are located around Barseek drain it is expected that high levels of nutrients and wastes concentrations reaches the lake directly as non point sources from this location. A scenario is developed to reduce the phosphorous concentrations at Barseek drain boundaries by 10% and 30% reductions, as shown in Figure (9-3).

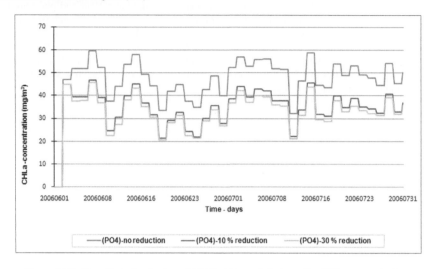

Figure (9-3): The effect of reduction of (PO4) at Barseek drain on the CHL-a concentrations at mid lake location

The reduction of PO4 at only at barseek drain outlet to the lake does not lower the CHL-aconcentration more than 10%, but reduction of nutrients from all input sources to the lake seems to be the reasonable scenario for reduction of primary production of algae in the lake. The scenarios results also show that the limiting factor to the CHL-a in the lake is both the phosphorous and nitrogen compounds. These results indicates that the management scenarios for the reduction of nutrients loads to the lake needs intensive reduction of nutrients in the watershed upstream and the 10% reduction of loads at drains inlets to the lake is a the starting management control action. Further research and analysis should be done on spatial and temporal variation of CHL-a concentrations based on the reduction scenarios to select the best management scenario taking into consideration all the surrounding factors.

10.   CONCLUSIONS AND RECOMMENDATIONS

This research explored the surface water quality problems in irrigated watersheds that comprises connected complex drainage networks and shallow lakes systems, and developed a computational framework for better understanding of the hydrodynamic and water quality processes within these systems. The methodologies used depended on the integration between different existing tools for management involving, mathematical models, GIS and Remote sensing. This chapter presents the conclusions drawn from the research work and highlights the main research contributions, and the recommendations for future research and development in the field of surface water quality management using combined and integrated techniques and data sets.

10.1. CONCLUSIONS

10.1.1 Surface Water Quality Management in Data Scarce Environments

Increasing attention is being paid to the management of water resources on a watershed basis; this implies a cross-disciplinary approach to definition of problems, data collection and analysis. The assessment of surface water quality on a watershed scale, involves the examination of all activities in the watershed for their possible effects on the existing water bodies. This in turn demands the existence of various temporal and spatial categories of data. Irrigated agricultural watersheds having a complex physical basis such as interacting water bodies e.g. canals, drains and coastal lagoons need a holistic management approach. In the 'data scarce environments' there is usually a lack of essential historical, regularly monitored and measured water quality datasets to study the water quality problems and develop the appropriate tools. In such data poor environments the problems of managing water quality becomes more obvious and the need for reliable solutions becomes an urgent requirement. Therefore this research work has focused on the management of surface water quality problems in such watersheds and the importance of taking into consideration all the watershed components and the effects of pollution from the upstream canals on the downstream coastal lakes. The detailed research study also emphasised the important need for reassessment and integration of available different information technology tools designed to support the management process.

10.1.2 Framework for Integrating Tools to Develop Water Quality Management Information System (WQMIS)

An approach for developing a framework for integrating tools for surface water quality management is presented in this research study. This framework makes use of the existing tools that includes GIS, remote sensing, hydrodynamic, water quality and eutrophication modelling tools. This framework uses the simple technology of geo-database capabilities of GIS which structures the different types of inputs that are needed in the modelling process. It also includes a set of models that are applied on the site specific conditions of the Edko watershed and shallow lake system for better understanding of the system hydrodynamics and water quality processes. These models are integrated with remote sensing extracted data sets for water quality that are needed for calibration and verification. An important result of this study is that under this framework the procedures for applying remote sensing analysis tools for calibration and verification of most important modelled water quality and eutrophication parameters are formulated and tested. The methodologies for linking between models and remote sensing to overcome the gaps in rare *in situ* measurements and the limited spatial variability of mathematical models are detailed in the study.

10.1.3 (1D-2D) Hydrodynamic Modelling of Catchment-Lake System (water balance)

Irrigated watersheds have a specific form, and they include a complex system of interacting surface water networks and open surface water bodies. Water quality models are considered key elements in understanding water quality problems for management and decision support processes. The understanding of water quality problems in such interacting systems needs a proper description and understanding of their hydrodynamic status. Therefore this research has emphasised the application of different levels of hydrodynamic models that better explains the catchment-lake system interactions and hydrodynamics properties. As a component of the framework for surface water quality management information system a (1D-2D) hydrodynamic model is developed to understand the basic hydrodynamics of the catchment–lake system. This model and gives an overview of the main parameters controlling the flow rates within the system. The 1D component represents the catchment part with its drainage network, and the 2D part links the catchment hydrodynamics to the lake to test and check the flow rates discharging to the lake water body, which is the focus of water quality modelling in this study. The model gives the basic information on how the system components are linked and what are their hydrodynamic characteristics under the different flow conditions, especially when the main boundary in the upstream catchment is a control point or pumping stations are controlling the flow within the main drainage system, and the downstream component is a shallow lake which is controlled by the flow coming from the catchment and the tidal condition at the sea exit channel. This model was viewed as important in order to understand the catchment-lake interactions.

10.1.4 Advanced 2D Shallow Lake Hydrodynamic Model

Since the shallow coastal lake at the downstream of the catchment is considered as a sink that receives all the wastewater discharged from the watershed, it was important to develop a more detailed modelling component for the lake system. The 2D hydrodynamic model was developed to facilitate a more detailed study of the lake hydrodynamics, taking into account the effects of the main driving forces on the flow which are wind and tides. The model was tested for different hydrodynamic scenarios to determine the most sensitive parameters that affect the flow conditions within the lake. The model showed that the main driving force that affects the flow velocities and currents in the lake is the wind force. The wind is responsible for mixing and resuspension in the lake due to its shallow depth, and this in turn is an important parameter to be considered during the water quality modelling of the lake system.

10.1.5 Lake Water Quality and Eutrophication Screening Models

A reliable water quality model is based on a detailed and well structured hydrodynamic model that is capable of describing the physical and hydrodynamic processes of the water system. Therefore the water quality and eutrophication modelling tools for the shallow lake system in this research work are coupled with the developed and calibrated hydrodynamic 2D model. The water quality modelling tools are the main components of the WQMIS. They were divided into two main models: the basic parameters model and the eutrophication screening model. The basic water quality modelling component simulates the main water quality parameters including the oxygen compounds (BOD, COD, DO), nutrients compounds (NH4, NO3, PO4), the temperature, salinity and the total suspended sediments (TSM). This model was able to predict the basic water quality indicators of the lake system. The second component is the eutrophication screening model for the lake. This is based on the simulation of CHL-aorophyll-a concentration, which is considered as an indicator of phytoplankton abundance and biomass in coastal and estuarine waters. Both models were applied on Edko shallow lake at the fringes of the western Delta region of Egypt. The models tested the effect of the pollution discharges from the catchment on the lake water quality. The basic parameters model showed high concentrations of BOD, COD and nutrients (NO3 and PO4) that exceed the standard limits. The eutrophication screening model showed high concentrations of TSM and CHL-a, which also exceeded the standards limits to reveal a high eutrophication of Lake Edko. The eutrophication screening model was also used to predict different management scenarios for the prevention of serious eutrophication and deterioration of the lake.

10.1.6 Use of RS for calibrating and verifying mathematical models for lake water quality parameters including TSM and CHL-a

Remote sensing is presented in this research study as an essential component in the WQMIS as it is used in different applications within the research. The first basic use of remote sensing is for data acquisition where ASTER and SPOT images were used to delineate the lake system components including the water body, the vegetation buffer zone and the fisheries areas. The main use of remote sensing data sets and the different associated analysis techniques was formulated in this research for the calibration of shallow lake water quality models focusing on TSM and CHL-a as the main parameters of concern. The combination of different types of remote sensing data (including MODIS09 and SPOT 5 images) provided the optimal setup between spectral and spatial resolution needed for the extraction of water quality parameters. The application of remote sensing in the calibration and verification processes of TSM and CHL-a for the Edko shallow lake gave reliable and successful results through the integration of remote sensing data with mathematical modelling in order to overcome the lack of *in situ* measurements and the limitation of real time data sets needed for the calibration of the models. Due to the high spatial, spectral and temporal capabilities of remote sensing data, it was possible to get further temporal and spatial variations of TSM and CHL-a concentrations within the lake from the images in order to compensate for the limited spatial variance of the mathematical models.

Since remote sensing technologies are advancing and the scientific community is rapidly improving the technical capabilities of satellites and processing tools, it is important that specific satellites are launched and used for specific research purposes connected with water quality modelling. A particular case in this point is the EVNISAT-MERIS data set which includes hyperspectral images captured from the EVNVISAT satellite of European Space Agency-ESA. These images are used specifically for water quality research. This satellite was launched in early 2003. It has been successfully used in this research work to verify the results obtained from SPOT-5 images. A major problem associated with MERIS is that it has a spatial resolution of 300m that does not give sufficient details in a case like Lake Edko. Therefore, there is a need for strong integration with higher resolution images to perform a detailed analysis. The integration of associated tools also implies the integration of different data types and information sources in order to obtain higher confidence levels in mathematical models.

This study is in line with new research approaches for the management of different environmental problems which focus on the integration of tools. This is shown by a similar approach taken by GEOSS (Global Earth System of Systems), which was an initiative of the global Earth Observation Secretariat in Geneva, which aims at linking and integrating *in-situ*, airborne and satellite data into one global homogeneous system for better management of critical problems such as environmental and climate change issues. Through this approach it is foreseen that the integration of data sets will give better quality analysis results and better understanding of systems under investigation.

10.2. RESEARCH ACHIEVEMENTS AND CONTRIBUTIONS

This research work has contributed mainly to surface water quality management. The main research hypothesis and concepts were based on the development of a computational framework that could be a part of a DSS for surface water quality management, integrating different modelling and analysis tools. The application of physically-based models needs reliable and available data sets in order to successfully include different variables, especially in water quality models. Therefore, the development of this framework as a methodology for managing water quality of shallow lakes is considered important in making better use of modelling capabilities in data scarce environments. The framework consisting of integrated tools has led to the development of better management approaches that are based on more certain knowledge.

The calibration of water quality models using remote sensing data and applying remote sensing analysis methodologies, is considered another important contribution of this research study. This research has shown that it is possible with acceptable statistical reliability to integrate mathematical water quality models and remote sensing extracted water quality parameters such as TSM and CHL-a for model calibration. Despite all the uncertainties involved with mathematical water quality models and with remote sensing analysis techniques, it has been possible to integrate both types of tools to strengthen the modelling system by bringing it closer to reality; this is because remote sensing is able to capture the real temporal status of the system under investigation.

The application of the developed WQMIS and its computational framework including the models and the remote sensing tools to a specific study area such as Edko catchment and lake system is another achievement of this research. The pollution threats to the northern coastal lakes in Egypt as the example of the lake Edko, call for the important need for the integration of different analytical and management tools. This integration helps to better understand the current status of the system and indicates the possible management scenarios. In particular, the developed computational framework for lake water quality management is considered an essential component of a decision support system for the management of a lake. The developed hydrodynamic, water quality and eutrophication models were able to give better understanding of the catchment and lake systems and they also succeeded in focussing on different water quality processes within the lake. The developed framework and the applied modelling system and tools on Lake Edko are generic and can be applied to other coastal lakes of the Nile Delta region or other regions with similar conditions.

10.3. RECOMMENDATIONS FOR FUTURE RESEARCH

This research study focused on two major issues: managing surface water on a watershed scale and the integration of management tools for deriving modelling systems with improved confidence for management purposes. For future continuation of research in this field the following recommendations are highlighted:

- This research was implemented originally to deal with data scarce environments, which is a common situation with many regions where there is a current lack of planning and management tools. But improvements in field monitoring in watersheds (or catchments) and shallow coastal lakes is still recommended with the field monitoring having a clear objective related to the modelling objectives and calibration needs. For water quality measurements and sampling the temporal and spatial frequency of measurements is critical. A specific monitoring plan should be developed to cope with the needs of both hydrodynamic and water quality models, and this plan should be consistent and linked with the remote sensing data collection.

- Improving the water balance model of the lake system taking into consideration the specific analysis and modelling of groundwater is recommended to be developed further in future research. Also needed is an improvement in the evapotranspiration rates within the water balance modelling as a key parameter in the water budget modelling. Evapotranspiration rates can be further studied with more focus on the rates associated with different vegetation species using remote sensing data and analysis techniques. Application of remote sensing in the calculation of evapotranspiration rates would be another added value of the integration of remote sensing with mathematical models; this in turn can contribute to better level of confidence.

- Remote sensing data sets should be refined for improved water quality and eutrophication modelling calibration. Since remote sensing was the main source of calibration data for water quality models in this research, it is recommended that more high resolution spatial and spectral data sets that are suitable for water quality analysis and extraction of water quality parameters are developed. An intensive frequency temporal data set of SPOT and MERIS images would be of considerable importance for further validation of the eutrophication modelling results.

- Attention should be given to the development of a site specific bio-optical remote sensing water quality model for the Delta region coastal lakes. In this present research the focus was on the mathematical modelling tools and their calibration using alternative data sets from remote sensing analysis. The remote sensing analysis methodologies used were based on empirical and statistical approaches, but for further studies of coastal lakes and monitoring their status, it is recommended to apply an advanced remote sensing analytical approach through the development of a site- specific bio-optical model that can be applied to these lakes. This model will require specific *in situ* measurements that can be done once and verified and calibrated every year. This model would be the key model for the application of remote sensing analysis and the extraction of the main parameters TSM and CHL-a. It could also be applied using any satellite imagery for water quality modelling.

- The inclusion of an advanced ecological model within the computational framework of the surface water quality management system is a recommended to be considered in the future. The current study has focused on the water quality modelling of basic parameters and the screening of eutrophication models. For better management of the lake water system it is important to focus on detailed ecological modelling, taking into consideration the detailed modelling of phytoplankton species, the modelling of the vegetation buffer zone around the water body and also the areas of fish ponds. The three components of the lake system explained in this research should be linked together in a more advanced and integrated hydrodynamic-ecological model. This will give access to more knowledge about the downstream area of the watershed, which could be modelled holistically as a wetland region.

- Conducting a detailed error propagation and uncertainty analysis that includes the different models and tools of the computational framework is highly recommended. Water quality and eutrophication modelling is associated with a high level of uncertainty due to the complex chemical and biological processes involved. This is in addition to the various possible sources of uncertainty that have been discussed in this study. Another important issue is that the use of remote sensing data to compensate for the lack of *in situ* measurements creates another accumulation of uncertainty associated with the application of the remote sensing. Therefore access to the acquisition of specific *in situ* measurements and more frequent remote sensing data can be used to explore the inherent uncertainties in order to improve model validation.

- Using the developed models for forecasting water quality is another important recommendation for future research. The modelling tools developed in this research and the integrated remote sensing tools have been mainly used to calibrate and verify models that are to be used for the prediction of water quality parameter concentrations. But these modelling and remote sensing tools could also be used for forecasting the water quality status of the lake system for any near future interventions within the watershed. The forecasting will depend on detailed remote sensing analytical tools to reduce the model uncertainty. Time series data sets could then be used for short term forecasting of TSM and CHL-a in the lake. Different forecasting methodologies could be applied and tested for best performance.

REFERENCES

Abbott, M. B. (1979). Computational hydraulics: Elements of the theory of free surface flows. Pitman, London.

Abdel-Dayem, S. (1994). Towards an Environmentally Sound Water Quality Management. Presented at the IWRA World Congress in Cairo.

Adriaanse, M., Van de Kraats, J., Stoks, P.G. and Ward, R.C. (1995a). Conclusions monitoring tailor-International Workshop Monitoring Tailor-made. Institute for Inland Water Management and Waste Water Treatment (RIZA), Lelystad, The Netherlands.

Adriaanse, M., H.A.G. Niederländer and P.B.M. Stortelder, (1995). Monitoring Water Quality in the Future, Chemical Monitoring. Institute for Inland Water Management and Waste Water
Treatment (RIZA), Lelystad. ISBN 90-802637-1-0.

Adriaanse, M. J., J. van der Kraats, P.G. Stoks and R.C. Ward (eds.), (1995). Proceedings of the international workshop Monitoring Tailor-Made I, 1994, Beekbergen, the Netherlands, 356pp.

Amer, M. H., Ridder, N. A. (1989). Land Drainage in Egypt. Ministry of Public Works and Water Resources, Cairo, Egypt.

Amos, C. L., Tplis, B.J. (1985). Discrimination of suspended particulate matter in the bay of Fundy using Nimbus 7 Coastal Zone Scanner. Can. J. Remote Sens. 11: pp.85-92.

Argent, R. M. a. R. B. G. (2001). "Design of information system for environmental managers: An example using interface prototyping." Environmental Modelling and Software 16(5): pp.433-438.

Arnbrose, R., Jr. (ed). (1992). Technical Guidance Manual for Performing Waste load Allocations, Book III - Estuaries - Part 4: Critical Review of Coastal Embayment and Estuarine Waste Load Allocation Modelling. Office of Water, US EPA, Washington, D.C. pp. 10-1 - 14-18.

Ascough, J. C., Maierb, H.R., Ravalico J.K., Strudleyc, b, M.W (2008). Future research challenges for incorporation of uncertainty in environmental and ecological decision-making. Ecological Modelling 219: pp. 383–399.

Aspinall, R. J., Marcus, A., W. , Boardman, J., W. (2002). Considerations in collecting, processing, and analysing high spatial resolution hyperspectral data for environmental investigations. J Geograph Syst 4: pp.15–29.

Bedford, K., Findikakis A., Larock B.E., Rodi W. and Street R.L. (1988). "Turbulence modeling of surface water flow and transport" J. Hyd. Eng. 114(9): 970-1073.

Bierman, V. J., Jr., and D., M, Dolan (1986). Modelling phytoplankton in Saginaw Bay I, Calibration phase. Journal of Environmental Engineering. 112(2): pp.400-413.

Branson, R. L., Pratt P. F., Rhoades J. D. and Oster J. D. (1975).Water Quality in Irrigated Watersheds."Journal of Environmental Quality 4: pp.33-40

Brooks, K. N, P.F. Ffolliott, H.M. Gregersen, and L.F. Debano. 1997. Hydrology and the Management of Watersheds, 2nd Ed. Iowa State University Press, Ames, Iowa.

Burges, S. J., Lettenmaier, D.P., (1975). "Probabilistic methods in stream quality management. ." Water Resources Bulletin 11 (1): pp.115–130.

Buttcher, D. (2003). Approaches for nutrient management in the lake okeechobee watershed, Symposium handbook, Rotorua lakes 2003 Practical management For good lake water quality, Newzeland.

Canale, R. P. a. S., D.I. (1996). Performance, Reliability and Uncertainty of Total Phosphorus Models for Lakes.2. Stochastic Analysis, Water Research, 30: pp.95-102.

Carpenter, D. S., Carpenter, S.M. (1983). Monitoring inland water quality using Landsat data.Remote Sensing Environ 13: pp.345-352.

C-CORE, 2007, Satellite Monitoring of Lake Water Quality in Egypt — Validation and Final Project Report (D46 and D50), C-CORE Report R-07-042-404 v1.1.

Chadderton, R. A., Miller, A.C., McDonnell, A.J., (1982). Uncertainty analysis of dissolved oxygen model. Journal of the Environmental Engineering Division 108 (EE5): pp.1003–1013.

Chaoa, X., Jia, Y., Shields , F., D.,Wanga,S. Y. S., Cooper, C. M., (2007). Numerical modeling of water quality and sediment related processes. Ecological Modelling 201: pp.385–397.

Chapman, D., Ed. (1996). Water quality assessments. A guide to the use of biota, sediments, and water in environmental monitoring, UNESCO/WHO/UNEP. Chapman &Hall.

Chapra, S. (1997). Surface water quality modeling. MacGraw Hill, New York.

Chapra, S. C. (2003). Engineering Water Quality Models and TMDLs. Journal of Water Resources Planning and Management, ASCE Vol. 129(Issue 4): pp.245-355.

Chapra, S. C. (1999) Organic carbon and surface water quality modelling). Prog. Environ. Sci. pp: 49–70.

Chen, C. W., and Orlob, G.T., (1975). Ecological simulation of aquatic environments. Systems analysis and Simulation in Ecology vol. 3 ed. B C. Patten (New York Academic Press) pp.476-588.

Choi, H. T., Beven, K., (2007). Multi-period and multi-criteria model conditioning to reduce prediction uncertainty in an application of TOPMODEL within the GLUE framework. Journal of Hydrology 332: pp.316–336.

Chubey, V. K., Subramanian, V. (1992). Estimation of suspended solids using Indian Remote Sending Satellite – I A data. : a case study from central India. International Journal of Remote Sensing 13: pp.1473-1486

Cieniawski, S. E., Eheart, J.W., Ranjithan, S., (1995). "Using genetic algorithms to solve a multiobjective groundwater monitoring problem." Water Resources Research 31(2): 399–409.

Ciu, L. J., Kuczera, G., (2005). "Optimizing water supply headworks operating rules under stochastic inputs: assessment of genetic algorithm performance " Water Resources Research 41, W05016, doi:10.1029/2004WR003517.

Cofino, W. P. (1995). Quality management of monitoring programmes. In: M. Adriaanse, J. Van de Kraats, P.G. Stoks and R.C. Ward [Eds] Proceedings of the International Workshop Monitoring Tailor-made. Institute for Inland Water Management and Waste Water Treatment (RIZA), Lelystad, The Netherlands.

Collins, C. D. (1988). Evaluating water quality for lake management. Final report. Technical Report, PB-89-148159/XAB. New York State Museum, Albany, NY (USA).

Cullen, P. (1990). The turbulent boundary between water science and water management. Freshwater Biology 24: pp.201-209.

Curran, P. J., Novo, E.M.M. (1988). The relationship between suspended sediment concentration and remotely sensed spectral radiance. A review. J. Coastal Res. 4: pp.351-368.

Curran, P. J. a. S., C. M. (2005). MERIS: The Re-branding of an Ocean Sensor. International Journal of Remote Sensing v. 26: pp. 1781-1798.

Dahl, M., Wilson, D., Hakansonc, L. (2006). A combined suspended particle and phosphorus water quality model: Application to Lake Vanern. Ecological Modelling 190: pp.55–71.

Dekker, A. G., Malthus, T.J., Wijnen, M.W., Seyhan, E. (1992). The effect of spectral bandwidth and positioning on the spectral signature analysis of inland water. Remote Sensing Environ 41: pp.211-225.

Dekker, A. G., Peters, S.W.M. (1993). The use of thematic mapper for the analysis of eutrophic lakes: a case study in the Netherlands. International Journal of Remote Sensing 14: pp. 779-821.

Demayo, A. a. S., A. (1996). Data handling and presentation. In: D. Chapman [Ed.] Water Quality Assessments. A Guide to the Use of Biota and E. Monitoring., Published on behalf of UNESCO, WHO and UNEP by Chapman & Hall, London 511-612.

Dilks, D. W., SC. Hinq RB. Ambrose, Ir., and JL. Martm (1990). Simplified Illustrative Examples In: Martin, J.M, R.B. Ambrose, Jr., and S.C. McCutcheon. Technical Guidance Manual for Performing Waste load Allocations, Book III - Estuaries. Part 2: Application of Estuarine Waste Load Allocation Models. US EPA (EPA 823/R-92-003) 6: pp.16-47.

Dunn, I. G. (1985). Aquatic weed control in relation to fisheries of lake Edku and Barseek fish farm, food and agriculture organization of the united nations (FAO), Rome.

Eheart, J. W., Ng, T.L., (2004). Role of effluent permit trading in total maximum daily load programs: overview and uncertainty and reliability implications. Journal of Environmental Engineering 130 (6): pp.615–621.

El-Sarraf, W. M. (1976). Some limnological studies on the submerged macrophytes in lake Edku, with special reference to their value as food of fish. MSc. Thesis. Fac. Sci. Alex. Univ. 117 pp.

Engman, E. T., and Gurney, R.J. (1991). Remote sensing in hydrology. (London:Chapman and Hall).

EPA, U. S. E. P. A. (2000). National Water Quality Inventory Report.

FAO (2000c). Agriculture towards 2015/30. Technical Interim Report, April 2000, Global Perspectives Study Unit.

FAO (2000a). Small ponds make a big difference. Integrating fish with crop and livestock farming. 30 pp.

Farinha, J., Costa LT., Zalidis, GC., Mantzavelas AL., Fitoka EN., Hecker N., Vives, TP. (1996). Mediterranean wetland Inventory: Habitat Description System MedWet/IUCN. Publicat Vol. No. IV.

Fedra, K. (1995). "Decision support for natural resources management: Models, GIS and expert systems. A.I. Applications, 9/3 (1995) pp 3-19."

Ferrier, G. (1995). A field study of the variability in suspended sediment concentration –reflectance relationship. International Journal of Remote Sensing 16: pp.2713-2720.

Froidefond, J. M., Castaing, P., Jouanneau, J.M., Prud'homme, R., Dinet, A. (1993). Method for the quantification of suspended sediments from AVHRR NOAA-11 satellite data. International Journal of Remote Sensing 14: pp.885-894.

Ganoulis, J. (1994). Engineering Risk Analysis of Water Pollution. Probabilities and fuzzy Sets. VCH Verlagsgesellschaft Weinheim, Germany.

Gardner, R. H., O'Neill, R. V., Mankin, J. B. & Kumar, D. (1980). Comparative error analysis of six predator prey models. Ecology 6(2), 323–332 6(2): pp.323–332.

Gitelson, A. A. (1992). The peak near 700 nm on radiance spectra of algae and water: relationships of its magnitude and position with CHL-aorophyll concentration. International Journal of Remote Sensing, 13: pp. 3367-3373.

Goodchild, M.F., (1996). Environmental modelling with GIS. New York: Oxford Univ. Press.

Gordon, H. G., Clark, D.K., Brown, O.B., Evans, R.H., Broenkow, W.W. (1983). Phytoplankton pigment concentrations, in the Middle Atlantic Bight: a comparison of ship determinations and CZCS estimates. Appl. Optics 22: pp. 20-35.

Gordon, H. R., Morel, A.Y., (1983). Remote assessment of ocean colour for interpretation of satellite visible imagery: A review. New York: Springer-Verlag. 114pp.

Grane'li, W. D. S. (1988). Influence of aquatic macrophytes on phosphorus cycling in lakes. Hydrobiologia 170: pp.245–266.

Haagsma, U. G. (1995). The integration of computer models and databases into a decision support system for water resources management. In S.P. Simonovic et al. (ed). Modelling and management of sustainable basin-scale water resources systems. IAHS. Publ. 231:pp. 253-261.

Haith, D. A., and Tubbs, L. J. (1981). Watershed loading functions for nonpoint sources. J. Envir. Eng. Div., ASCE, 107(1), 121–137. Environmental Protection Agency, Washington, D.C.

Harding, L. W., Itsweire, E.C., Esaias, W.E. (1995). "Algorithm development for recovering CHL-aorophyll concentrations in Chesapeak Bay using aircraft remote sensing." Photogrammetric Engineering Remote Sensing 61: pp. 177-185.

Harrington, J., J.A., Scheibe, F.R., Nix, J.F. (1992). "Remote sensing of lake Chicot, Arkansa.; Monitoring suspended sediments, turbidity and secchi depth with Land Sat and MSS " Remote Sensing Environ 39: 15-27.

Hartigan, J. P., Friedman, J. A. & Southerland, E. (1983). "Post-audit of lake model used for NPS management. J. Environ. Engng. (ASCE) 109(6): pp.1354–1370.

Hashimoto, T., Stedinger, J.R., Loucks, D.P., (1982). Reliability, resiliency, and vulnerability criteria for water resource system performance evaluation.Water Resources Research 18 (1): pp. 14–20.

Havnø, K., Madsen, M. N. & Drge, J. , Ed. (1995). MIKE 11—A Generalized River Modelling Package. In Computer Models of Watershed Hydrology (ed. V. P. Singh), pp. 733–782. Water Resources Publications.

Hem, J. D. (1985). Study and Interpretation of the Chemical Characteristics of Natural Water. U. S. Geological Survey Water-Supply Paper 2: pp.254-263.

Henderson-Sellers, B., and R. I. Davies,, Ed. (1991). Chapter 5: Model Validation and Sensitivity: Case Studies in the Global Context in Water Quality Modeling Volume IV: Decision Support Techniques for Lakes and Reservoirs (Ed. by Henderson-Sellers and French). , CRC Press, Inc. Florida.
.
Hogenboom, H. J. a. A. G. D. (1999). Report: Inversion: assessment of water composition from spectral reflectance. A feasibility study to the use of matrix inversion method. Delft, Report of the Netherlands Remote Sensing Board (BCRS) pp. 98-15.

Holyer, R. J. (1978). Toward universal multispectral suspended sediment algorithms. Remote Sensing and Environment, 7: pp. 323-338.

Horne, A. J. a. C. R. G. (1994). LAKE ECOLOGY OVERVIEW. Limnology. 2nd edition. N. Y. McGraw-Hill Co., New York, USA.

Hovis, W. A., Ed. (1981). The NIMBUS 7 coastal zone scanner. IN Gower, J.F.R. (Ed.) Oceanography from space, New York: Plenum Press pp. 213-226.

Hu, C., Chena, Z., Claytonb, T.D., Swarzenskib,P., and J. C. Brockb, & Muller-Karger, F. E. (2004). "Assessment of estuarine water-quality indicators using MODIS medium-resolution bands: Initial results from Tampa Bay, FL." Remote Sensing of Environment 93: pp. 423-441.

Hutchinson, G. E. (1957). A Treatise on Limnology. L Geography, Physics and Chemistry., John Wiley and Sons, New York,

Imam, H. E., and Kamal A. Ibrahim, (1996a). Minimum Nile Drainage Needs for Sustainable Estuarine Ecosystem, Special Study Report, NWRC of MPWWR and Winrock International Institute for Agricultural Development, US Agency for International Development, Cairo, Egypt.

Ivanov, P., Masliev I., Kularathna M., Kuzmin A., Somlyody L. (1995). DESERT: Decision Support System for Evaluating River Basin strategies. International Institute for Applied Systems Analysis (IIASA) Working Paper WP-95-23.

Jaeger, C., Renn, O., Rosa, E.A., Webler, T., (2001). Risk, Uncertainty and Rational Action. Earthscan Publications Ltd., London, 320 pp.

Jeppsson, U. (1996). Modelling Aspects of Wastewater Treatment Processes, PhD dissertation, Lund Institute of Technology (LTH), Department of Industrial Electrical Engineering and Automation (IEA).

Jerome, J. H., Bukata, R.P., Miller, J.R. (1996). Remote sensing reflectance and its relationship to optical properties of natural water. International Journal of Remote Sensing 15: pp. 53-62.

Joniak, T., A. Kamin´ ska & R. Gołdyn, (2000). Influence of the Wind on Vertical Changes of Water Properties in a Shallow Reservoir. In Gurgul, H. (ed.). Physicochemical Problems of Natural Waters Ecology, Scient. Publ. Univ, Szczecin. Vol. II.: pp.67–76.

Joniak, T., Kuczyn´ ska-Kippen, N., Nagengast, B., (2007). The role of aquatic macrophytes in microhabitatual transformation of physical-chemical features of small water bodies. Hydrobiologia 584: pp.101–109.

Jorgensen, S. E. (1976). An Eutrophic model for a lake. Ecological Modeling 2: pp.147-165.

Jorgensen, S. E. (1983). Ecological modeling of lakes. In Orlob, G.T., Mathematical modelling of water quality : Streams, lakes and reservoirs. John Wiley & Sons, New York, ISBN 047-1100315.

Jorgensen, S. E., Kamp-Nielsen, L., Christensen, T., Windolf-Nielsen, J. & Westergaard, B. (1986). "Validation of a prognosis based upon a eutrophication model. ." Ecol. Model. 32: pp.165–182.

Jupp, D., Kirk, J. and Harris, G. (1994). Detection, identification and mapping of cyanobacteria: using remote sensing to measure the optical quality of turbid inland waters. Aust. J. Mar. Fresh Res. 45: pp.801–828.

Kaden, S., A. Backer & A. Gnauck (1989). Decision support system for water management. In D.P. Loucks and U. S. (eds), Closing the gap between theory and practice. IAHS Publ. 180: pp.11-21.

Karaska, M. A., Huguenin, R.L., Beacham, J.L., Wang, M-H, Jensen, J.R., Kaufmann, R.S., (2004). AVIRIS Measurements of CHL-aorophyll, suspended minerals, dissolved organic carbon, and turbidity in the Neuse River, North Carolina. Photogram. Eng. and Remote Sens. 70(1): pp.125-133.

Karr, J. R., Duddly, D.R. (1981). Ecological perspectives on water quality goals. Environmental Management. 5: pp.55-68.

Karr, J. R., Fausch, K.D., Angermeier, P.L., Yant, P.R., & SCHL-aosser, I.J., (1986). Assessing biological integrity in running waters: a method and its rationale. Special Publ. 5. Illinois Natural History survey.

Keller, P., I. Keller and K. I. Itten (1998). Combined hyperspectral data analysis of an Alpine lake using CASI and DAIS 7915 imagery. 1st EARSeL Workshop on Imaging Spectroscopy, Zürich.

Khedr, A. a. L.-D. J. (2000). Determenants of floristic diversity and vegetation composition on the islands of Lake Burollos, Egypt. App Veg Sci 3: pp.147–156.

Khorram, S. (1985). Development of water quality models applicable though the entire San Fransisco Bay and delta. Photogrammetric Engineering Remote Sensing 51: pp.53-62.

Kim, K., and Ventura, S. (1993). Large scale modeling of urban nonpoint source pollution using a geographical information system. Phtotogrametric Engineering and Remote Sensing 59(10): pp.1539-1544.

Kirk, J. T. O. (1984). Dependence on the relationship between inherent and apparent optical properties on solar altitude. Limnology and Oceanography 29: pp.350-356.

Kirk, J. T. O. (1994). Light and photosynthesis in aqatic ecosystems, Cambridge University Press.

Koponen, S., Pulliainen, J., Kallio, K., & Hallikainen, M. (2001). Lake water quality classification with airborne hyperspectral spectrometer and simulated MERIS data. Remote Sensing of Environment 79: pp.51-59.

Krijgsman, J. (1994). Optical propersties of Water Quality parameters. Interpretation of Reflectance Spectra, PhD thesis, Delft University of Technology: 200p.

Kritikos, H., Yorinks, L., Smith, H. (1974). Suspended solids analysis using ERTS-data. Remote Sensing Environ. 3: pp. 69-78.

Labiosa, W., Leckie, J., Shachter, R., Freyberg, D., Rytuba, J., (2005). Incorporating uncertainty in watershed management decision-making: a mercury TMDL case study. In: ASCE Watershed Management Conference, Managing Watersheds for Human and Natural Impacts: Engineering, Ecological, and Economic Challenges, Williamsburg, VA, July 19–22, doi:10.1061/40763(178)125.

Lathrop, J., R.G. Lillesand T.M. (1989). Monitoring Water quality and river plume transport I Green Bay, Lake Mitchigan with SPOT-1 imagery. Photogrammetric Engineering Remote Sensing 55: pp.349-354.

Lathrop, R. G. a. T. M. L. (1986). Use of Thematic Mapper data to assess water quality in Green Bay and central Lake Michigan. Photogrammetric Engineering & Remote Sensing, 52(5): pp.671-680.

Lopes, L., F., G., Do Carmob, J., S., A., Cortes, R., M., V., Oliveira D., (2004). Hydrodynamics and water quality modelling in a regulated river segment: application on the instream flow definition. Ecological Modelling (173): pp.197-218.

Lung, W. S., Larson, C.,E. (1995). Water Qaulity Modelling of Upper Mississippi River and Lake Pepin. Journal of Environmental Engineering. 121(10): pp. 691-699.

Maasdam, R. (2000). Exploratory Data Analysis in Water Quality Monitoring Systems, University of Salford. MSc. Thesis.

Magnus Dahl and David Wilson. Current status of freshwater quality models. Technical report, Karlstad University, May 2000.

Maguire, L. A., Boiney, L.G., (1994). Resolving environmental disputes—a framework incorporating decision analysis and dispute resolution techniques. Journal of Environmental Management 42 (1): pp.31-48.

Mahmood, K., Yevjevich, V. (1975). Unsteady flow in open channels, Vol. 1. Water Resources Publications, Littleton, CO, USA.

Mantovani, J. E., Cabral, A.P. (1992). Tank depth determination for water radiometric measurements. International Journal of Remote Sensing, 13: pp.2727-2733

Masato, N., Solomatine D.P., Nishida, W. (2002). Calibration of Water Qaulity Model by Global Optimisation Techniques\. Proc. 5th International Conference in Hydroinformatics, Cardiff, UK.

Mattikali, N. M. D., B.J., and Richards, K.S. (1996). Prediction of river discharges and surface water quality using an integrated geographical information system approach. International Journal of Remote Sensing 17(4): pp.683-701.

Mattikalli, N. M., and Engman, E.T. (2000). Integration of remotely sensed data into GIS: Remote sensing in hydrology and water management. Springer-Verlag, Berlin Heidelberg, ISBN 3-540-64075 4.

Mauersberger, P., Ed. (1983). General principles in deterministic water quality modeling, in G. T. Orlob (ed.), Mathematical Modelling of Water Quality: Streams, Lakes, and Reservoirs, John Wiley and Sons.

Mauersberger, P. (1983). General principles in deterministic water quality modeling, in G. T. Orlob (ed.), Mathematical Modelling of Water Quality: Streams, Lakes, and Reservoirs, John Wiley and Sons.

Mayo, M., Kamieli, A., Gitelson, A., Ben-Avraham, Z. (1993). Determination of suspended sediment concentrations from CZCS data. Photogrammetric Engineering Remote Sensing 59:pp.1265-1269.

McCutcheon, Steve C. (1990). Water Quality Modeling – Volume 1 – Transport and Surface Exchange in Rivers, Boca Raton: CRC Press Inc.

McIntyre R. Neil , W. T., Wheater, S.H. and Yu, S. Z. (2003). Uncertainty and risk in water quality modelling and management. Journal of Hydroinformatics 5: pp.259-274.

Meybeck, M., Chapman, D. and Helmer, R. [Eds] (1989). Global Freshwater Quality. A First Assessment. Blackwell Reference, Oxford, 306 pp.

Meybeck, M., Helmer, R. (1996). AN INTRODUCTION TO WATER QUALITY. Water Quality Assessments - A Guide to Use of Biota, Sediments and Water inbEnvironmental Monitoring - Second Edition. D. Chapman, UNESCO/WHO/UNEP.

Meybeck, M. a. R. H. (1989). The Quality of Rivers:from Pristine Stage to Global Pollution. Palaeogeography, Palaeoclimatology, Palaeoecology 75: pp.283–309.

Miettinen, K., M. (1998). Nonlinear Multiobjective Optimization, International Series in Operations Research & Management Science, 12, 1st Edition, Kluwer Academic Publishers, Boston, USA.

Miller, R. L. a. M., B.A. (2004). Using MODIS Terra 250 m imagery to map concentrations of total suspended matter in caostal waters. Remote sensing of Environment 93 (1-2):pp. 259-266.

Moldan, B., S. Billharz, and R. Matrazers, eds. , Ed. (1997). Sustainabillity Indicators SCOPE 58, Paris, France.

Monson, B. 1992. A primer on limnology, second edition. Water Resources Center, University of Minnesota, 1500 Cleveland Avenue, St. Paul, MN 55108, USA.

Morel, A. a. L. P. (1977). Analysis of variations in ocean colour. Limnology and Oceanography 22: pp.709-722.

Moss, B. (1988). Ecology of Fresh Waters, 2nd Ed. Man & Medium, Blackwell Scientific, Oxford.

Nandalal, K. D. W., S. P. Simonovic (2002). State-of-the-Art Report on Systems Analysis Methods for Resolution of Conflicts in Water Resources Management, Division of Water Sciences, UNESCO, Paris.

Naot, D., Rodi, W. (1982). Calculation of secondary currents in channel flow. J. Hyd. Div., 108(HY8): pp.948-968.

Novo, E. M. L. M., C.A. Stefen, and C.A. Braga, (1991). Results of a laboratory experiment of relating spectral reflectance to total suspended solids. Remote Sensing of Environment, 36:pp.57-72.

Novo, E. M. L. M. B., C. C. ; Freitas, R. M. ; Shimabukuro, Y. E. ; Melack, J. M. ; Pereira Filho,W. (2006). Seasonal changes in CHL-aorophyll distributions in Amazon floodplain lakes derived from MODIS images. Japanese Journal of Limnology, v1 : pp. 1-9.

Novo, E. M. M., Hansom, J.D., P.J. Curran, P.J. (1989a). The effect of sediment type on the relationship between reflectance and suspended sediment concentration. International Journal of Remote Sensing, 10: pp.1283-1289.

Ongley, E. D. (1993). Global Water Pollution: Challenges and Opportunities. In. Integrated Measures to Overcome Barriers to Minimize Harmful Fluxes from Land to Water. Proceedings of the Stockholm Water Symposium, Stockholm Water Company, Sweden.

Ongley, E. D. (1995). The global water quality programme. In: M. Adriaanse, J. Van de Kraats, P.O. Stoks and R.C. Ward [Eds] Proceedings of the International Workshop Monitoring Tailor-made. Institute for Inland Water Management and Waste Water Treatment (RIZA), Lelystad, The Netherlands.

Ongley, E. D. (2000). Water quality management: design, financing and sustainability considerations-II. Invited presentation at the World Bank's Water Week Conference: Towards a Strategy for Managing Water Quality Management, April 3-4, 2000, Washington, D.C. USA.

Orlob, G. T. (1983). Mathematical modelling of water quality,: Streams, lakes and reservoirs. John wiley & Sons, New York, ISBN 047-1100315.

Osman, E., M., Badr, B., E. (2010). Lake Edku water-quality monitoring, analysis and management model for optimising drainage water treatment using a genetic algorithm. Int. J. Environment and Waste Management, Vol. 5, Nos. 1/2.

Ottens, J. J., F.A.M. Claessen, P.G. Stoks, J.G. Timmerman, R.C Ward (eds.), (1997). Proceedings of the international workshop Monitoring Tailor-Made II, 1996, Nunspeet, the Netherlands, 492pp.

Pallottino, S., Secchi, G.M., Zuddas, P., (2005). A DSS for water resources management under uncertainty by scenario analysis. Environmental Modelling & Software 20 (8): pp.1031–1042.

Park, R. A. (1974). A generalized model to simulate lake ecosystems simulation

Parker, D. E., Ed. (1993). Climatic variation. In Concise Encyclopaedia of Environmental Systems (ed. P. C. Young), pp.102–111. Pergamon, Oxford.

Parmenter, B. M. (1991). The northern lakes of Egypt: Encounters with a wetland environment Austin, TX (US). Univ. of Texas PhD.

Peters, N. E., Meybeck, M. (2000). Water Quality Degradation Effects on Freshwater Availability: Impacts of Human Activities. IWRA, Water International Volume 25(2): pp.185–193.

Pierson, D. C. (1998). Measurement and modeling of radiance reflection in Swedish waters. 1st EARSeL Workshop on Imaging Spectroscopy, Zürich.

Prato, T. (2005.). Bayesian adaptive management of ecosystems. Ecological Modelling 183 (2-3): pp.147-156.

Prein, M., Ahmed, M. (2000). Integration of aquaculture into smallholder farming systems for improved food security and household nutrition. . Food Nutr. Bull 21: pp.466-471.

Pulliainen, J., Vepsalainen, J., Kallio, K., Koponen, S., Pyhalahti, T., Harma, P., & and M. Hallikainen, (n.d.). Monitoring of water quality in lake and coastal regions using simulated ENVISAT and MERIS data. Helsinki, Finland: Helsinki niversity of Technology, Laboratory of Space Technology.

Quibell, G. (1992). Estimating CHL-aorophyll concentrations using upwelling radiance from different freshwater algal generations. International Journal of Remote Sensing 13(2611-2621).

Radwan, M. (2002). River water quality modeling as water resources management tool at catchment scale, Ph.D. thesis, KU Leuven, Belgium

Radwan, M., Willems, P., and Berlamont, J. (2004). Sensitivity and uncertainty analysis for river quality modelling. Journal of Hydroinformatics 6(83-99).

Rafailidis, S., Ganoulis J. (1994). Impact of climatic Changes on coastal Water Quality. In: Angelakis A. (ed.) Diachronic climate Impacts on Water Resources. , NATO-ARW, Springer Verlag, Berlin-Heidelberg-New York.

Rao, M. (1998). Water quality monitoring of Blood River embayment – Kentucky Lake using field spectrometer and remote sensing studies.

Rardin, R., L. (1997). Optimization in Operations Research, Prentice Hall, Upper Saddle River, NJ, USA.

Reckhow, K. H. (1994). A decision analytic framework for environmental analysis and simulation modeling. Environmental Toxicology and Chemistry 13 (12): pp.1901–1906.

Reeves, P., De Winton, M., (2005). Fresh water biodiversity: Restoring lake vegetation. Water & Atmosphere 13(3).

Reichert, P., Borsuk, M.E., (2005). Does high forecast uncertainty preclude effective decision support. Environmental Modelling & Software 20 (8): pp.991–1001.

Rickert, D. (1993.). Water quality assessment to determine the nature and extent of water pollution by agriculture and related activities. In: Prevention of Water Pollution by Agriculture and Related Activities. Proceedings of the FAO Expert Consultation. Santiago, Chile, 20-23 October, 1992. Water Report 1. FAO, Rome. : pp. 171-194.

Rios Insua, D. (1990). Sensitivity Analysis in Multi-objective Decision Making. Springer-Verlag, Berlin, 193 pp.

Ritchei, J. C., Cooper, C.M., Scheibe, F. R. (1990). The relationship of multi-spectral scanner (MSS) and thematic mapper TM digital data with suspended sediments, CHL-aorophyll and temperature in Moon Lake, Mississippi. Remote Sensing Environ 33: pp.137-148.

Ritchie, J. C., Cooper, C.M. (1991). An algorithm for using Landsat MSS for estimating surface suspended sediments. Water Res. Bulletin 27: pp.373-379

Ritchie, J. C., Shciebe, F.R., Mc Henry, J.R. (1976). Remote sensing of suspended sediments in surface water. Photogrammetric Engineering Remote Sensing 42: pp.1539-1545.

Ritchie, J. C., and Scheibe, F.R., (2000). Water Quality: Remote sensing in hydrology and water management. Springer-Verlag Berlin Heidelberg, ISBN 3-540-64075-4 .

Ritchie, J. C. and C. M. Cooper (2002). Remote sensing techniques for determining water quality: application to TMDLs.

Rodi, W. (1993). Turbulence models and their application in hydraulics, 3rd ed., IAHR/AIRH monograph, Balkema, Rotterdam.

Rosenthal, R., E. (1985). Concepts, Theory, and Techniques – Principles of Multiobjective Optimization. Decision Sciences Vol. 16: pp.133-152.

Rundquist, D. C., Han, L. Schalles, J.F., Peake, J.S. (1996). Remote measurement of algal CHL-aorophyll in surface waters. The case of the first derivative of reflectance near 690nm. Photogrammetric Engineering Remote Sensing 62: pp.195-200

Sabherwal, R., V. Grover (1989). "Computer Support for Strategic Decision-Making Processes: Review and Analysis." Decision Sciences Vol. 20: pp.54-76.

Sage, A. P., Armstrong, J. E., (2000). Introduction to Systems Engineering, Wiley-Interscience, New York.

Sagehashi, M., , Sakoda, A., Suzuki, M. (2001). A mathematical model of a shallow and eutrophic lake (the keszthely basin, lake balaton) and simulation of restorative manipulations. wat. res. vol. 35 (no. 7): pp. 1675–1686.

Samaan, A. A. (1974). Primary production of the Edku Lake, Egypt Bull Inst Ocean and Fish, ARE 4: pp.260–317.

Scheffer, M. (2001). Alternative attractors of shallow lakes. The Scientific World 1 254-263.

Schiller, D. R. a. H. (2008). MERIS Regional Coastal and Lake Case 2 Water Project - Atmospheric Correction ATBD. GKSS Research Center 21502 Geesthacht Version 1.0 18. May 2008.

Shafique, N., A., Fulk, F., Autrey, B., C., and Flotemersch J. (2002). Hyperspectral Remote Sensing of Water Quality Parameters for Large Rivers in the Ohio River Basin.

Shafique, N. A., Autrey, B. C , Fulk, F., & Cormier, S. M. (2001). The selection of narrow wavebands for optimizing water quality monitoring on the Great Miami River, Ohio using hyperspectral remote sensing data. Journal of Spatial Hydrology.

Shepherd, B., Harper, D. & Millington, A. (1999). Modelling catchment-scale nutrient transport to watercourses in the UK. Hydrobiologia 396: pp.227–237.

Shultz, G. A. (1988). Remote sensing in hydrology. Journal of Hydrology 100: pp.239-265.

Somlyody, L., Henze, M., Koncsos, L., Rauch, W., Reichert, P., Shanahan, P. & Vanrolleghem, P. (1998). River water quality modelling: III. Future of the art. Wat. Sci. Technol 38(11): pp.253-260.

Stephan, G. H., Fang X. (1993). Model Simulations of Dissolved Oxygen Characteristics of Minnesota Lakes: Past and Future. Environmental Management Vol. 18(1): pp. 73-92.

Strumpf, R. P., Tayler, M.A. (1988). Satellite detection of bloom and pigment distributions in estuaries. Remote Sensing Environ 24: pp.385-404.

Szpakowska, B. (1999). Occurrence and Role of Organic Compounds Dissolved in Surface and Ground Waters of Agricultural Landscape, Nicolai Copernici University Press, Torun´

Tang, Y. A. (1977). Report of the Aquaculture Mission in Egypt,, Rome, FAO, 48p.

Texas Water Development Board (1970a). DOSAGE1 Simulation of water quality in streams and canals. Program documentation and user's manual.

Texas Water Development Board (1970b). QUAL1 Simulation of water quality in streams and canals. Program documentation and user's manual.

Thomann, R. (1982). Verification of Water Quality Models. Journal of the Environmental Engineering Division 108(EE5): pp.923-940.

Thomann, R., Mueller, J.A (1987). Pxinciples of Surface water Quality Modelling and ControL, Harper Collins Publishers, Inc., New York 644 p.

Thomann, R. V. (1998). The future "golden age" of predictive models for surface water quality and ecosystem management. . J. Environ. Engng. (ASCE) 124(2): pp.94–103.

Thomann, R. V. (1998). The future 'Golden Age' of predictive models for surface water quality and ecosystem management. J. Environ. Eng. 124(2): pp.94–103.

Thomann, R. V. (1972). Systems analysis and water quality management, McGraw-Hill, New York.

Thomas, R. M. M. a. B. A. (1996). Lakes. Water Quality Assessments - A Guide to Use of Biota, Sediments and Water in Environmental Monitoring -Second Edition. D. Chapman, UNESCO/WHO/UNEP.

Timmerman, J.G., J.J. Ottens and R.C. Ward, 2000. The information cycle as a framework for defining information goals for water-quality monitoring. Environmental management 25 (3): pp.229 - 239.

United Nations Commission for Sustainable Development (1997). Comprehensive Assessment of the Fresh Water Resources of the World. Geneva, Switzerland: World Meteorological Organization.

Van der Perk, M. (1997). Effect of model structure on the accuracy and uncertainty of results from water quality models. Hydrol. Process 11 (3): pp.227–239.

Van Puijenbroek, P. J. T. M., Janse,J.H. , Knoop, J.M. (2004). Integrated modelling for nutrient loading and ecology of lakes in The Netherlands. Ecological Modelling 174: pp.127–141.

Vasquez, J. A., Maier, H.R., Lence, B.J., Tolson, B.A., Foschi, R.O., (2000).Achieving water quality system reliability using genetic algorithms. Journal of Environmental Engineering, ASCE 126(10): pp.954–962.

Vollenweider, R. A. (1968). Scientific Fundamentals of the Eutrophication of Lakes and Flowing Waters with Special Reference to Nitrogen and Phosphorus as Factors in Eutrophication. Technical Report DA5/SCI/68.27, Organisation for Economic Cooperation and Development, Paris, 250 pp.

Vollenweider, R. A. (1968). Scientific Fundamentals of the Eutrophication of Lakes and Flowing Waters, with Particular Reference to Nitrogen and Phosphorus as Factors in Eutrophication. Organisation for Economic Co-operation and Development, Paris.

Ward, R. C., J.C Loftis and G.B. McBride, (1986). The "data-rich but information-poor" syndrome in water quality monitoring. Environmental Management 10(3): pp.291-297.

Ward, R. C. (1995a). Monitoring tailor-made: what do you want to know? In: M. Adriaanse, J. Van de Kraats, P.G. Stoks and R.C. Ward [Eds] Proceedings of the International Workshop Monitoring Tailor-made. Institute for Inland Water Management and Waste Water Treatment (RIZA), Lelystad, The Netherlands.

Water Resources Engineers, I. (1973). Computer program documentation of the stream quality model QUAL II. Report to US Environmental Protection Agency, Washington, DC.

Whitlock, C. H., Kuo, C. Y., Le Croy, S.R. (1982). Criteria for the use of regression analysis for remote sensing of sediment and pollutants. Remote Sensing Environ. 12: pp.151-168.

World Bank, (1998). Pollution Prevention and Abatement Handbook.

World Bank. (2000). World Development Report 2000/2001: Attacking Poverty.

Yang, M. D., Kuo, J. T and Yeh F. Y.(2000). Application of Remote Sensing and GIS in Water Quality Simulation and Calibration. 4th International Conference on Integrating GIS and Environmental Modelling (GIS/EM4): Problems, Prospects and Research Needs.

Young, P., Parkinson, S. & Lees, M.(1996). Simplicity out of complexity in environmental modelling: Occam's razor revisited. J. Appl. Statist. 23(2): pp.165–210.

Zacharias, I., Gianni, A., (2008). "Hydrodynamic and dispersion modeling as a tool for restoration of coastal ecosystems. Application to a re-flooded lagoon." Environmental Modelling & Software. 23: pp.751-767.

APPENDIX (A-1)

Moderate-resolution Imaging Spectroradiometer (MODIS) and the SPOT-5 Satellite Images Characteristics

MODIS (Moderate-resolution Imaging Spectroradiometer)

MODIS is an extensive program using sensors on two satellites that each provide complete daily coverage of the earth. The data have a variety of resolutions; spectral, spatial and temporal. The MODIS web site, http://modis.gsfc.nasa.gov/index.php, is a good place to begin learning about this important program. This site has links to the Atmospheres, Land and Oceans groups of MODIS. Because the MODIS sensor is carried on both the Terra and Aqua satellites, it is generally possible to obtain images in the morning (Terra) and the afternoon (Aqua) for any particular location. Night time data are also available in the thermal range of the spectrum. You should consider time of day when ordering a scene for a specific day.

Daily MODIS Scenes – Level 1

Individual daily MODIS scenes, MODIS *Level 1* products, can be obtained for any part of the earth, every day, since February 2000. These files are in the Geographic projection and have a spatial resolution of 1 km. There are 36 bands of image data interspersed through 85 total bands of information in the file. The data are digital numbers in *16 bit unsigned integer* format. These data should be converted to radiance values, surface reflectance values, and/or brightness/temperature values before performing any analysis. Terra file names for the complete file begin with MOD021KM. Aqua file names begin with MYD021KM.

Alternatively daily scenes at 500 m resolution can be obtained which would include the first 7 bands of data, or the first two bands of data at 250 m resolution. The Terra names for these are MOD02HKM and MOD02QKM respectively.

MODIS Products - Level 2

Product data are generally provided in the sinusoidal projection. Below are descriptions of the surface reflectance and vegetation products, but many more products are available.

MODIS Surface Reflectance Products; MOD 09 - Surface Reflectance; Atmospheric Correction Algorithm Products

The MODIS Surface-Reflectance Product (MOD 09) is computed from the MODIS Level 1B land bands 1, 2, 3, 4, 5, 6, and 7 (centered at 648 nm, 858 nm, 470 nm, 555 nm, 1240 nm, 1640 nm, and 2130 nm, respectively). The product is an estimate of the surface spectral reflectance for each band as it would have been measured at ground level if there were no atmospheric scattering or absorption. The correction scheme includes corrections for the effect of atmospheric gases, aerosols, and thin cirrus clouds; it is applied to all noncloudy MOD 35 Level 1B pixels that pass the Level 1B quality control. The correction uses band 26 to detect cirrus cloud, water vapor from MOD 05, aerosol from MOD 04, and ozone from MOD 07; best-available climatology is used if the MODIS water vapor, aerosol, or ozone products are unavailable. Also, the correction uses MOD 43, BRDF without topography, from the previous 16-day time period for the atmosphereBRDF coupling term. The following are examples of the

MOD09GQK - MODIS Surface Reflectance Daily L2G Global 250m

This file has a spatial resolution of 250 m and contains two bands of spectral data centered at 645 nm and 858 nm. There are also three bands of additional information on band quality, orbit and coverage, and number of observations.

MOD09Q1 - MODIS Surface Reflectance 8-Day L3 Global 250m

This file is a composite using eight consecutive daily 250 m images. The "best" observation during each eight day period, for every cell in the image, is retained. This helps reduce or eliminate clouds from a scene. The file contains the same spectral information as the daily file listed above, centered at 645 nm and 858 nm. There is one additional band of data for quality control.

MOD09GHK - MODIS Surface Reflectance Daily L2G Global 500m

This file has a spatial resolution of 500 m and contains seven bands of spectral data plus three bands of additional information on band quality, orbit and coverage, and number of observations. The spectral range for each band can be found in Appendix A, bands 1 through 7.

MOD09A1 - MODIS Surface Reflectance 8-Day L3 Global 500m

This file is a composite using eight consecutive daily 500 m images. The "best" observation during each eight day period, for every cell in the image, is retained. This helps reduce or eliminate clouds from a scene. The file contains the same seven spectral bands of data as the daily file listed above. It also has an additional 6 bands of information concerning quality control, solar zenith, view zenith, relative azimuth, surface reflectance 500 m state flags, and surface reflectance day of year.

MOD09GQK (*Used in this research Study*)

MODIS Terra/Aqua Surface Reflectance Daily L2G Global 250 m

Product description:

MOD09GQK provides MODIS band 1-2 daily surface reflectance at 250 m resolution. The best observations during a 24-hour period, as determined by overall pixel quality and observational coverage, are matched geographically according to corresponding 250 m Pointer Files. Quality information for this product is provided at three different levels of detail: for individual pixels, for each band and each resolution, and for the whole file.

Science Data Sets for MOD09GQK

Science Data Sets (HDF Layers) (5)	Units	Data Type	Fill Value	Valid Range	Scale Factor
250m Surface Reflectance Band 1 (620-670 nm)	Reflectance	16-bit signed integer	-28672	-100 - 16000	0.0001
250m Surface Reflectance Band 2 (841-876 nm)	Reflectance	16-bit signed integer	-28672	-100 - 16000	0.0001
250m Reflectance Band Quality (*see Table 12*)	Bit field	16-bit unsigned integer	2995	NA	NA
Orbit and coverage (*see Table 20*)	Bit field	8-bit unsigned integer	15	0 - 255	NA
Number of Observations	none	8-bit signed integer	-1	0 - 127	NA

Web site: http://modis-sr.ltdri.org

SPOT-5

SPOT (Système Probatoire d'Observation de la Terre) (lit. "Probationary System of Earth Observation") is a high-resolution, optical imaging Earth observation satellite system operating from space. It is run by Spot Image based in Toulouse, France. It was initiated by the CNES (Centre national d'études spatiales — the French space agency) in the 1970s and was developed in association with the SSTC (Belgian scientific, technical and cultural services) and the Swedish National Space Board(SNSB). It has been designed to improve the knowledge and management of the Earth by exploring the Earth's resources, detecting and forecasting phenomena involving climatology and oceanography, and monitoring human activities and natural phenomena. The SPOT system includes a series of satellites and ground control resources for satellite control and programming, image production, and distribution. The satellites were launched with the ESA rocket launcher Ariane 2, 3, and 4. The company SPOT Image is marketing the high-resolution images, which SPOT can take from every corner of the Earth.

SPOT 1 launched February 22, 1986 with 10 panchromatic and 20 meter multispectral picture resolution capability. Withdrawn December 31, 1990.

SPOT 2 launched January 22, 1990 and deorbited in July 2009.

SPOT 3 launched September 26, 1993. Stopped functioning November 14, 1997

SPOT 4 launched March 24, 1998

SPOT 5 launched May 4, 2002 with 2.5 m, 5 m and 10 m capability

SPOT 5 was launched on May 4, 2002 and has the goal to ensure continuity of services for customers and to improve the quality of data and images by anticipating changes in market requirements. SPOT 5 has two high resolution geometrical (HRG) instruments that were deduced from the HRVIR of SPOT 4. They offer a higher resolution of 2.5 to 5 meters in panchromatic mode and 10 meters in multispectral mode (20 metre on short wave infrared 1.58 - 1.75 μm). SPOT 5 also features an HRS imaging instrument operating in panchromatic mode.

SPOT-5 multispectral sensor

Description

Satellite orbital details as per SPOTs 1-4.(60-80 km swath; 26-day orbital repeat cycle for nadir viewing; generally 11 viewing opportunities every 26 days over New Zealand).The spatial resolution of SPOT-5 multispectral data is 10 m. This means that SPOT-5 bridges the gap between the medium resolution services and the very high resolution services.

Availability

SPOT-5 was launched on 4 May 2002. Data is downloaded into a number of receiving stations overseas. The nearest station to New Zealand is in Adelaide and its coverage circle cannot reach N.Z. Therefore, N.Z. data is only available via the on-board tape recorders.

Uses

Mapping and monitoring for agriculture, forestry, exploration, land use, geology, hydrology, coastal resources, coastal monitoring, ship detection and monitoring. Suitable for mapping scales down to at least 1:25 000 scale.

SPOT satellite spectral bands and resolutions

sensor	electromagnetic spectrum	pixel size	spectral bands
SPOT 5	Panchromatic	2.5 m or 5 m	0.48 - 0.71 µm
	B1 : green	10 m	0.50 - 0.59 µm
	B2 : red	10 m	0.61 - 0.68 µm
	B3 : near infrared	10 m	0.78 - 0.89 µm
	B4 : mid infrared (MIR)	20 m	1.58 - 1.75 µm
SPOT 4	Monospectral	10 m	0.61 - 0.68 µm
	B1 : green	20 m	0.50 - 0.59 µm
	B2 : red	20 m	0.61 - 0.68 µm
	B3 : near infrared	20 m	0.78 - 0.89 µm
	B4 : mid infrared (MIR)	20 m	1.58 - 1.75 µm
SPOT 1 SPOT 2 SPOT 3	Panchromatic	10 m	0.50 - 0.73 µm
	B1 : green	20 m	0.50 - 0.59 µm
	B2 : red	20 m	0.61 - 0.68 µm
	B3 : near infrared	20 m	0.78 - 0.89 µm

From raw images to digital satellite image maps

SPOT images come with different levels of geometric preprocessing, divided into 2 product lines:

SPOT Scene (*Used in this research study*)

level 1A: Radiometric correction of distortions due to differences in sensitivity of the elementary detectors of the viewing instrument. Intended for users who wish to do their own geometric image processing.

level 1B: Radiometric correction identical to that of level 1A. Geometric correction of systematic effects (panoramic effect, Earth curvature and rotation). Internal distortions of the image are corrected for measuring distances, angles and surface areas. Specially designed product for photo interpreting and thematic studies.

level 2A: Radiometric correction identical to that of level 1A. Geometrical correction done in a standard cartographic projection (UTM WGS84 by default) not tied to ground control points. Allowing for possible differences in location, this product is used to combine the image with geographical information of various types (vectors, raster maps and other satellite images).

Image format

The default SPOT Scene delivery format is DIMAP SPOT Scene profile. The only exceptions are 2.5-metre and 5-metre colour SPOT Scene 2A products, which are delivered in DIMAP SPOTView profile format.

SPOTView

level 2B (Precision): This product comes in a map projection with ground control points taken on maps or from GPS type measurements taken in the field. The image is corrected for a mean elevation in a projection and a standard map frame. This product is used when deformations due to relief are not that important (flat ground, etc.).

level 3 (Ortho): Map projection based on ground control points and a DEM based on Reference3D data to eliminate distortions due to relief.

Image format

SPOTViews are delivered in DIMAP SPOTView profile format.

APPENDIX (A-2)

Atmospheric Correction of Satellite Images

The nature of remote sensing requires that solar radiation pass through the atmosphere before it is collected by the instrument. Because of this, remotely sensed images include information about the atmosphere and the earth's surface. For those interested in quantitative analysis of surface reflectance which is the case for water quality parameters, removing the influence of the atmosphere is a critical pre-processing step. To compensate for atmospheric effects, properties such as the amount of water vapour, distribution of aerosols, and scene visibility must be known. Because direct measurements of these atmospheric properties are rarely available, there are techniques that infer them from their imprint on hyperspectral radiance data. These properties are then used to constrain highly accurate models of atmospheric radiation transfer to produce an estimate of the true surface reflectance. Moreover, atmospheric corrections of this type can be applied on a pixel-by-pixel basis because each pixel in a hyperspectral image contains an independent measurement of atmospheric water vapor absorption bands.

ENVI's Fast Line-of-sight Atmospheric Analysis of Spectral Hypercubes (FLAASH) module is a first-principles atmospheric correction modeling tool for retrieving spectral reflectance from hyperspectral radiance images. With FLAASH, we can accurately compensate for atmospheric effects. FLAASH corrects wavelengths in the visible through near-infrared and short-wave infrared regions, up to 3 mm. (For thermal regions, use the Basic Tools Preprocessing Calibration Utilities Thermal Atm Correction menu option.) Unlike many other atmospheric correction programs that interpolate radiation transfer properties from a pre-calculated database of modeling results, FLAASH incorporates the MODTRAN4 radiation transfer code.

We can choose any of the standard MODTRAN model atmospheres and aerosol types to represent the scene; a unique MODTRAN solution is computed for each image. FLAASH also includes the following features: Correction for the adjacency effect (pixel mixing due to scattering of surface-reflected radiance). An option to compute a scene-average visibility (aerosol/haze amount). FLAASH uses the most advanced techniques for handling particularly stressing atmospheric conditions, such as the presence of clouds.

FLAASH supports hyperspectral sensors (such as HyMAP, AVIRIS, HYDICE, HYPERION, Probe-1, CASI, and AISA) and multispectral sensors (such as Landsat, SPOT, IRS, and ASTER). Water vapor and aerosol retrieval are only possible when the image contains bands in appropriate wavelength positions (see Input Data Requirements for details). In addition, FLAASH can correct images collected in either vertical (nadir) or slant-viewing geometries.
FLAASH was developed by Spectral Sciences, Inc., a world leader in optical phenomenology research, under the sponsorship of the U.S. Air Force Research Laboratory. Spectral Sciences has been an integral part in the development of modern atmospheric radiation transfer models, and has worked extensively on MODTRAN since the model's inception in 1989.

The ENVI FLAASH Model

This section is a brief overview of the atmospheric correction method used by FLAASH.
FLAASH starts from a standard equation for spectral radiance at a sensor pixel, L, that applies to the solar wavelength range (thermal emission is neglected) and flat, Lambertian materials or their equivalents. The equation (1) is as follows:

$$L = \left(\frac{A\rho}{1 - \rho_e S}\right) + \left(\frac{B\rho_e}{1 - \rho_e S}\right) + L_a$$

Where:

ρ is the pixel surface reflectance

ρ e is an average surface reflectance for the pixel and a surrounding region

S is the spherical albedo of the atmosphere

La is the radiance back scattered by the atmosphere

A and B are coefficients that depend on atmospheric and geometric conditions but not on the surface.

Each of these variables depends on the spectral channel; the wavelength index has been omitted for simplicity. The first term in Equation (1) corresponds to radiance that is reflected from the surface and travels directly into the sensor, while the second term corresponds to radiance from the surface that is scattered by the atmosphere into the sensor. The distinction between r and re accounts for the adjacency effect (spatial mixing of radiance among nearby pixels) caused by atmospheric scattering. To ignore the adjacency effect correction, set re = r. However, this correction can result in significant reflectance errors at short wavelengths, especially under hazy conditions and when strong contrasts occur among the materials in the scene.

The values of A, B, S and La are determined from MODTRAN4 calculations that use the viewing and solar angles and the mean surface elevation of the measurement, and they assume a certain model atmosphere, aerosol type, and visible range. The values of A, B, S and La are strongly dependent on the water vapor column amount, which is generally not well known and may vary across the scene. To account for unknown and variable column water vapour, the MODTRAN4 calculations are looped over a series of different column amounts, then selected wavelength channels of the image are analyzed to retrieve an estimated amount for each pixel. Specifically, radiance averages are gathered for two sets of channels: an absorption set cantered at a water band (typically 1130 nm) and a reference set of channels taken from just outside the band. A lookup table for retrieving the water vapor from these radiances is constructed. For images that do not contain bands in the appropriate wavelength positions to support water retrieval (for example, Landsat or SPOT), the column water vapor amount is determined by the user-selected atmospheric mode.

After the water retrieval is performed, Equation (1) is solved for the pixel surface reflectances in all of the sensor channels. The solution method involves computing a spatially averaged radiance image Le, from which the spatially averaged reflectance re is estimated using the approximate equation (2):

$$L_e \approx \left(\frac{(A + B)\rho_e}{1 - \rho_e S}\right) + L_a$$

Spatial averaging is performed using a point-spread function that describes the relative contributions to the pixel radiance from points on the ground at different distances from the direct line of sight. For accurate results, cloud-containing pixels must be removed prior to averaging. The cloudy pixels are found using a combination of brightness, band ratio, and water vapour tests, as described by Matthew et al. (2000).

The FLAASH model includes a method for retrieving an estimated aerosol/haze amount from selected dark land pixels in the scene. The method is based on observations by Kaufman et al. (1997) of a nearly fixed ratio between the reflectances for such pixels at 660 nm and 2100 nm. FLAASH retrieves the aerosol amount by iterating Equations (1) and (2) over a series of visible ranges, for example, 17 km to 200 km. For each visible range, it retrieves the scene-average 660 nm and 2100 nm reflectances for the dark pixels, and it interpolates the best estimate of the visible range by matching the ratio to the average ratio of ~0.45 that was observed by Kaufman et al. (1997). Using this visible range estimate, FLAASH performs a second and final MODTRAN4 calculation loop over water.

218

APPENDIX (A-3)

ENVISAT Medium-Resolution Imaging Spectrometer (MERIS)

Launched in 2002, ENVISAT is the largest Earth Observation spacecraft ever built. It carries ten sophisticated optical and radar instruments to provide continuous observation and monitoring of the Earth's land, atmosphere, oceans and ice caps. Envisat data collectively provide a wealth of information on the workings of the Earth system, including insights into factors contributing to climate change. The Medium Resolution Imaging Spectrometer (MERIS) is a imaging spectrometer that measures the solar radiation reflected by the Earth, at a ground spatial resolution of 300 m, with 15 spectral bands in visible and near infra-red and programmable in width and position. MERIS allows global coverage of the Earth every 3 days.

The primary mission of MERIS is the measurement of sea colour in oceans and coastal areas. Knowledge of sea colour can be converted into a measurement of CHL-aorophyll pigment concentration, suspended sediment concentration and aerosol loads over marine areas. It is also used for land and atmospheric monitoring.

MERIS products
The MERIS Level 1 products provide the radiance as measured by the instrument. The can be used to construct colour composite images, either giving a photographic impression or highlighting certain thematic features of the Earth, and they are input to processing algorithms to derive Level 2 products. The MERIS level 2 products provide geophysical information ready to be used for various applications. The primary mission of MERIS is to monitor the ocean colour including CHL-aorophyll concentrations for open oceans and coastal areas, yellow total suspended matter. In addition, MERIS provides with land parameter measurements like vegetation indices and atmospheric parameters like water vapour, cloud top pressure, cloud types, aerosol optical thickness, and Angström coefficients.

MERIS Full Resolution products

Instrument / mode	Product ID	Description
	MER_FR__1P	Full Resolution Geolocated and Calibrated TOA Radiance
	MER_FR__2P	Full Resolution Geophysical Product for Ocean, Land and Atmosphere
FRS	MER_FRS_1P	Full Resolution Full Swath Geolocated and Calibrated TOA Radiance
	MER_FRS_BP	Full Resolution Full Swath Browse product
	MER_FRS_2P	Full Resolution Full Swath Geophysical Product for Ocean, Land and Atmosphere

Image size of MERIS products

Product	image size	ground coverage
Full Resolution Full scene	2241 pixels x 2241 lines	582 km (across-track) by 650 km (along-track)
Full Resolution Quarter scene	1153 pixels x 1153 lines	300 km (across-track) by 334 km (along-track)

219

MERIS Image Bands and Spectral Characteristics

Band	Wavelength [nm]	Width [nm]	Potential Applications
1	412.5	10	Yellow substance, turbidity
2	442.5	10	Chlorophyll absorption maximum
3	490	10	Chlorophyll, other pigments
4	510	10	Turbidity, suspended sediment, red tides
5	560	10	Chlorophyll reference, suspended sediment
6	620	10	Suspended sediment
7	665	10	Chlorophyll absorption
8	681.25	7.5	Chlorophyll fluorescence
9	705	10	Atmospheric correction, red edge
10	753.75	7.5	Oxygen absorption reference
11	760	2.5	Oxygen absorption R-branch
12	775	15	Aerosols, vegetation
13	865	20	Aerosols corrections over ocean
14	890	10	Water vapor absorption reference
15	900	10	Water vapor absorption, vegetation

Samenvatting

Belangrijke problemen die het gevolg zijn van specifieke aspecten van menselijk samenleven, tasten de waterkwaliteit aan van rivieren, beken en meren. Problemen komen voort uit onvoldoende behandeld afvalwater, slechte manieren van landgebruik, ontoereikende controles op de afvoer van industriële afvalwaterstromen, verkeerde locaties van industriële terreinen, onbeheerste en schamele landbouw praktijken, buitensporig gebruik van meststoffen en een tekort aan geïntegreerd management van het stroomgebied. De resultaten van deze problemen bedreigen de ecosystemen, brengen openbare gezondheidszorgrisico's met zich mee en verhevigen erosie en sedimentatie, leidend tot kwaliteitsvermindering van land en waterbronnen. Vele van deze negatieve resultaten komen voort uit milieu vernietigende ontwikkelingen, een tekort aan informatie over de toestand van de waterkwaliteit en gebrekkig publiek bewustzijn en scholing gericht op de bescherming van water bronnen. Geïrrigeerde stroomgebieden zijn bijzonder gevoelig voor zulke waterkwaliteitsproblemen.

Toenemende aandacht wordt besteed aan het watermanagement vanuit de optiek van het stroomgebied, hetgeen een benadering noodzakelijk maakt van de definitie van problemen, de gegevensverzameling en de analyse die dwars door alle disciplines heen gaat. De beoordeling van oppervlaktewaterkwaliteit op de schaal van een stroomgebied, betrekt in haar onderzoek alle activiteiten in het stroomgebied op hun mogelijke effecten op de bestaande waterlichamen. Geïrrigeerde stroomgebieden zijn dikwijls complex in hun fysieke natuur doordat zij op elkaar inwerkende irrigatie en drainage netwerken omvatten, die aangesloten kunnen zijn op meren of lagunes. Het bestuderen van waterkwaliteitsproblemen in zulke stroomgebieden ten behoeve van beter beheer vraagt om een herwaardering en integratie van informatietechnologische hulpmiddelen, die ontworpen zijn om het beheerproces te ondersteunen.

Waterkwaliteitsmodellen worden beschouwd als hoofdonderdelen voor het begrijpen van waterkwaliteitsproblemen en ze zijn belangrijke componenten in beheer en beslissingsondersteunende systemen. Modellen zijn tegenwoordig zeer geavanceerd in het beschrijven van de dynamica van de aquatische omgeving en kunnen een aanzienlijke hoeveelheid gegevens produceren, die moeilijk zijn te begrijpen. Het probleem dat zich dikwijls voordoet is het selecteren van de meest efficiënte manier om die gegevens in hun geografische context voor te stellen. Aan de andere kant is de geografische verwerking van milieu-informatie goed ontwikkeld en vele geavanceerde Geografische Informatiesystemen (GIS) zijn nu verkrijgbaar. Ook van aardobservatie technieken is in diverse studies aangetoond dat zij goede mogelijkheden bezitten voor het in kaart brengen en controleren van een aantal parameters van waterkwaliteit. Daarom zou de integratie van wiskundige modellen, GIS en aardobservatie- toepassingen een krachtig hulpmiddel kunnen bieden voor management en besluitvormingsprocessen die betrekking hebben op waterkwaliteitsproblemen. Maar ook biedt deze integratie een belangrijke benadering om het probleem van schaarste aan gegevens in zulke omgevingen te overwinnen.

Het onderzoek dat in deze studie wordt gepresenteerd is gericht op het bijdragen aan het veld van management van de oppervlaktewaterkwaliteit door het integreren van wiskundige waterkwaliteitsmodellen op basis van fysische kennis van het systeem, mét de ruimtelijke mogelijkheden van GIS én de ruimtelijke en temporele mogelijkheden van aardobservatie ("remote sensing") teneinde een geïntegreerd management-informatie-systeem voor waterkwaliteit te ontwikkelen dat toepasbaar is voor geïrrigeerde stroomgebieden.

Het succes van een geïntegreerde benadering van het beheer van waterkwaliteit in een stroomgebied is sterk afhankelijk van de beschikbaarheid van gegevens. Zulke gegevens verschillen in vorm al naar gelang de verschillende niveaus van gebruikers, zoals bijvoorbeeld voor bestuurders en beleidsplanners of voor de technische ingenieurs die in het veld werken. Een essentieel hulpmiddel dat gebruikt wordt door waterkwaliteitsonderzoekers en managers bij het ontwikkelen van beheerplannen voor rivieren, beken en meren, is wiskundig modeleren. De integratie van verschillende computertechnologieën en hulpmiddelen, zoals GIS en de snel groeiende aardobservatie technologie als een krachtige gegevensbron, geeft samen met waterkwaliteitsmodellen een zelfs nog krachtiger en efficiënt beheerinstrument, zeker wanneer men te maken heeft met gecompliceerde waterafvoer-netwerken in stroomgebieden.

Binnen deze onderzoeksstudie is een algemeen kader ontwikkeld voor een informatiesysteem om het beheer van oppervlaktewaterkwaliteit te faciliteren. Dit systeem is gebaseerd op de integratie van hydrodynamische en waterkwaliteit modellen met GIS en teledetectie als instrumenten voor het genereren van beheer scenario's voor oppervlaktewaterkwaliteit in een geïrrigeerd stroomgebied. De toepassing is ontwikkeld voor het Edko afwateringsgebied en het ondiep meersysteem in het westelijke deel van de Nijl Delta, Egypte. Het kader omvat een hiërarchie van modelleer hulpmiddelen: een 1D-2D basis hydrodynamisch model voor een gecombineerd ondiep meer-afwateringssysteem, een gedetailleerd 2D hydrodynamisch model van het ondiepe meer, en een 2D waterkwaliteits en eutrofiërings model voor het meer-systeem. Bovenop deze modelleer hulpmiddelen worden aardobservatie-gegevens gebruikt om de wiskundige waterkwaliteitsmodellen te kalibreren en te valideren.

Als onderdeel van het informatiesysteem van de waterkwaliteitsbewaking voor het Edko afwateringsgebied en het ondiepe meer, is een 1D-2D hydrodynamisch model ontwikkeld om de basale hydrodynamica van het stroomgebied van een meer-systeem te begrijpen. Voor verdere analyse van de waterkwaliteit en eutrofiërings gesteldheid van het meer, is een meer gedetailleerd 2D hydrodynamisch-waterkwaliteitsmodel ontwikkeld. Dit model is gebaseerd op het 1D-2D modelleringconcept met een nadruk op hoofdparameters die de hydrodynamica van een meer beïnvloeden zoals de wind en de getijde-tijdreeks, de verdampingsverliezen van het waterlichaam en de evapotranspiratie van aquatische vegetatie. Betrouwbare waterkwaliteitsmodellen zijn gebaseerd op gedetailleerde en goed gestructureerde hydrodynamische modellen die de fysieke en hydrodynamische processen van het watersysteem kunnen beschrijven. Overmatige nutriëntenbelasting die tot eutrofiëring leidt, is een gemeenschappelijk probleem in de meeste ondiepe meren, vooral voor diegenen die gekoppeld zijn aan landbouw afwateringssystemen. Daarom waren gedetailleerde modellen van de waterkwaliteit voor het meersysteem vereist om opgenomen te worden in het kader. 2D hydrodynamische en specifieke waterkwaliteits- en eutrophiërings-modellen werden ontwikkeld voor het aan de kust gelegen ondiepe meer Edko.

Het model van de waterkwaliteit heeft allereerst een component die de belangrijkste parameters van de waterkwaliteit simuleert met inbegrip van de zuurstofsamenstellingen (BOD, COD, DO), nutrienten samenstellingselementen (NH4, NO3, PO4), de temperatuur, het zoutgehalte en de totale opgeloste stof (TSM, total suspended matter). Het model voorspelt de basisindicatoren van de waterkwaliteit van het meersysteem. De tweede component is het eutrophiërings onderzoeksmodel voor het meer; dit is gebaseerd op de simulatie van de CHL-aorofyll-a concentratie die wordt beschouwd als een indicator van fytoplankton-overvloed en biomassa in kust en estuariene wateren. Een succesvolle toepassing van een model vereist een kalibratie die een vergelijking omvat van de gesimuleerde resultaten met de gemeten gesteldheid van het meer, waarbij gebruik gemaakt wordt van verschillende kalibratiehulpmiddelen. De kalibratie van de waterkwaliteitsmodellen is gedaan door gebruik te maken van verschillende technieken om zeker te zijn van vergelijkbare uitkomsten van het model met de gemeten gegevens. Het model is aanvankelijk gekalibreerd door de parameters aan te passen voor geselecteerde waterkwaliteitsprocessen.

Een tweede niveau van model-kalibratie was gericht op de TSM geproduceerde concentratiepatronen welke voorkomen uit aardobservatie-gegevens, die op hun beurt verkregen zijn uit de analyse en het verwerken van een tijdreeks gegevensset van MODIS beelden (MOD09), specifiek voor tijdelijke kwalitatieve kalibratie. Een gedetailleerde kwantitatieve ruimtelijke kalibratie werd gedaan voor concentraties TSM die gebaseerd zijn op beschikbare *in situ* metingen en de tijdreeksen van MODIS satellietbeelden, via het toepassen van TSM analysealgoritmen. De SPOT-5 Satelliet afbeelding werd gebruikt om er de concentraties CHL-a uit af te leiden teneinde het eutrophiëringsmodel te kalibreren; de resultaten werden geverifieerd door gebruik te maken van een MERIS-FR afbeelding. Het gebruik van aardobservatie-analyseprocedures en -gegevens voor de kalibratie en controle van de waterkwaliteits-wiskundige modellen in gegevens schaarse omgevingen heeft aangetoond dat het een waardevolle en betrouwbare benadering biedt.

Deze studie onderzocht de integratie van wiskundige modellen en aardobservatie methodieken en heeft aangetoond dat er waarde zit in het identificeren en het werken met de ruimtelijke en tijdgebonden variatie van de waterkwaliteitsparameters binnen het waterlichaam van het meer. De gekalibreerde modellen werden gebruikt om scenario's te ontwikkelen voor nutriëntenvermindering ten behoeve van het beheer van de kwaliteit van het meerwater. De verminderingsscenario's waren afhankelijk van het verlagen van de nutriëntenstroom van het stroomgebied naar het meer. Het onderzoek slaagde er in om de model-behoeften met verschillende bestaande hulpmiddelen te verbinden ten behoeve van beter beheer, waarbij rekening is gehouden met de praktische beperkingen, en het kiezen van een uitvoerbare en betrouwbare benadering voor het ontwikkelen van een kader om oppervlaktewaterkwaliteit te beheren in ondiepe meersystemen met geïrrigeerde stroomgebieden.

About the Author

Amel Moustafa Azab graduated at the Civil Engineering-Irrigation and Hydraulics Department, Faculty of Engineering, Cairo University, Egypt in 1996. She started her career after graduation as teacher assistant in fluid mechanics and hydraulics at the Civil Engineering Department of the American University in Cairo. In 1998 she joined the National Water Research Centre as research assistant at the Water Resources Research Institute, Hydrology department then Nile Basin Research department. Since 2000 she was appointed at the Hydraulics Research Institute as a research assistant and as a senior technical officer in the Nile Basin Capacity Building Network (NBCBN); a regional project of the UNESCO-IHE in collaboration with the Hydraulics Research Institute. In 2001 she received her Master of Science Degree in Water Resources Management from Cairo University, Civil Engineering-Irrigation and Hydraulics Department. By mid of 2005 she received a PhD research fellowship from UNESCO-IHE in the framework of the PoWER project. Her PhD research topic on surface water quality management is reported in this book. Being a PD research fellow at the Hydroinformatics Department of UNESCO-IHE was a part time position parallel to the her work at NBCBN, where her present position as the Manager of the Nile Basin Capacity Building Network. She is involved in various collaborative regional and international research activities and research projects. Her main research interest is in the field of wetlands and water quality management.

T - #0146 - 160425 - C238 - 240/170/13 - PB - 9780415621151 - Gloss Lamination